cocktail
party
guide to
GREEN
ENERGY

cocktail party guide to
GREEN ENERGY

everything you need to know to converse
intelligently about alternative energy

ANNETTE SALIKEN
with **MARTIN G. CLARKE**

HERITAGE

VANCOUVER | VICTORIA | CALGARY

Heritage House Publishing Company Ltd.
www.heritagehouse.ca

LIBRARY AND ARCHIVES CANADA CATALOGUING IN PUBLICATION

Saliken, Annette, 1961–
  Cocktail party guide to green energy: everything you need to know to converse intelligently about alternative energy / Annette Saliken, with Martin G. Clarke.

Includes bibliographical references and index.
Issued also in electronic format.
ISBN 978-1-926936-93-2

  1. Renewable energy sources—Popular works. I. Clarke, Martin G II. Title.
III. Title: Everything you need to know to converse intelligently about alternative energy.

TJ808.S24 2011   333.79'4   C2011-905039-0

Edited by Lesley Reynolds
Proofread by Karla Decker
Cover and book design by Jacqui Thomas
Cover photos by Imaj/iStockphoto.com (wind turbine), craftvision/iStockphoto.com
  (martini glass) and costint/iStockphoto.com (lightbulb)
Illustrations by Gary Fabbro and Annette Saliken except for:
  REN21. 2011. Renewables 2011 Global Status Report. Paris: REN21 Secretariat, 20
  and Dr. Colin J. Campbell of the Association for the Study of Peak Oil & Gas, 22

 The interior of this book was printed on 100% post-consumer recycled paper, processed chlorine free and printed with vegetable-based inks.

Heritage House acknowledges the financial support for its publishing program from the Government of Canada through the Canada Book Fund (CBF), Canada Council for the Arts and the province of British Columbia through the British Columbia Arts Council and the Book Publishing Tax Credit.

 Canadian Heritage Patrimoine canadien   The Canada Council | Le Conseil des Arts for the Arts | du Canada   BRITISH COLUMBIA ARTS COUNCIL

15 14 13 12 11   1 2 3 4 5

Printed in Canada

This book is
dedicated to
my husband,
MARTY,
who brings
both adventure
and serenity
into my life
every day.

CONTENTS

PREFACE

EASTER ISLAND, NOW WINDSWEPT, GRASSY and eerily quiet, was once home to tropical forests, vast bird colonies and thousands of human inhabitants. Today, stone statues erected by those islanders are scattered across this isolated and barren South Pacific island, facing out to sea as if whispering a ghostly warning to those passing by.

We now rely on archaeology to help us understand the day-to-day lives of Easter Island citizens hundreds of years ago. The island's inhabitants ate plentiful fruits from trees as well as birds that lived among those trees. Wood was harvested for warmth, cooking and building homes, canoes and tools. Forests were cleared to make way for crops and settlements. Eventually, the forests were gone and most tree species were extinct, which dramatically affected all aspects of human existence on the island. Without wood for canoes and spears, it became more difficult to venture out and catch much-needed seafood for the growing population. Even worse, with the decreasing supply of canoes it would have been difficult to move or escape from the island. While there is debate over the cause of the population collapse on Easter Island, many experts agree that overexploitation of natural resources played a role.

We can only wonder what was in the minds of those who cleared the last forests on Easter Island. Did they realize they were destroying a naturally renewable source of food, warmth, shelter, transportation and tools? Did they understand that this vital resource could last forever if they replanted and carefully managed their trees?

We do not know why the people of Easter Island decided to cut down so many trees. Maybe the decision was driven by economics—after all, the scarce timber must have become very valuable to the inhabitants of the island. Perhaps their priorities were short-term shelter or warmth, or cultural or religious needs.

The cautionary tale of Easter Island brings an important and urgent message for our society about sustainability, defined as "the ability to meet today's economic, environmental and social needs without compromising those needs of future generations." Our planet is now facing an exploding population and overconsumption of resources. We are already experiencing the consequences, such as a pending energy shortage and human-related climate change. We need to incorporate sustainability into our decision making in all aspects of our lives and at all levels—personal, business and government—to ensure the well-being of the planet for current and future generations.

By pursuing the renewable energy technologies described in this book, we can address our growing energy needs in a sustainable manner that moves our society toward innovation and economic growth and away from greenhouse-gas emissions, oil spills and rising fuel prices that will burden future generations.

INTRODUCTION

AS A FOLLOW-UP TO *Cocktail Party Guide to Global Warming*, which explains the fundamentals of climate change, this book provides readers with everything they need to know to understand green-energy options, carry on knowledgeable discussions and make informed energy choices in their everyday lives. It uses clear, objective language to explain alternatives to fossil fuel–based energy and describes how renewable energy can be implemented in homes and vehicles to reduce our carbon footprint.

The book explains why we need green energy and shares exciting breakthroughs in solar, wind, geothermal, geoexchange, hydro, wave, tidal and biomass energy, which offer environmentally friendly solutions to fuel shortages, global warming and energy security in our lifetime.

Also explored in these pages are clean-energy technologies that do not generate energy but make its delivery and storage more efficient. To enable readers to make informed decisions, the book compares the pros and cons as well as the costs of various types of green energy, answers frequently asked questions and clarifies common misconceptions.

Green energy is no longer a futuristic notion—it's rapidly entering the mainstream of society and has become a common topic in newspapers and magazines and on television and radio. In our own neighbourhoods, more developers are adopting greener energy options such as geoexchange heating and renewable district-energy systems. Meanwhile, car dealers are offering us more choices, such as high-efficiency, electric and hybrid vehicles. As a result, most of us are starting to face decisions that will affect our households and day-to-day lifestyles. What kind of energy do we want to use in our homes? What are the options? What are the pros and cons? Why does it matter? Growing environmental concerns, rising fuel prices and an increasing number of energy choices have resulted in the public seeking more information about green energy.

The chapters of this book are designed to be read either individually or in sequence and serve as educational references that allow readers to interpret everything they see, hear and read about green energy as we move into the future.

chapter one

WHY GREEN ENERGY?

Rising **FUEL PRICES,**

GLOBAL WARMING

and **OIL SPILLS** are

just a few reasons why

many people are now

seeking greener

energy options.

A CLOSER LOOK AT OUR ENERGY OPTIONS

There are many energy options to supply our day-to-day needs for cooking, heating, lighting, transportation and communications. Figure 1.1 shows a breakdown of energy sources into fossil fuel–based energy (including oil, coal, natural gas and their derivatives) versus alternative energy, which refers to alternatives to fossil fuel–based energy. There are two types of alternative energy: renewable and non-renewable. Renewable energy types, such as solar, wind and geothermal, are naturally replenished in nature, so they can continue indefinitely.

Nuclear energy is non-renewable since it draws on a limited resource, uranium, which is not replenished in nature. Although it is a relatively cost-effective and pollution-free source of electricity, nuclear energy could negatively affect future generations by posing a risk of nuclear accidents or unknown environmental and safety consequences from spent nuclear fuel rods. In

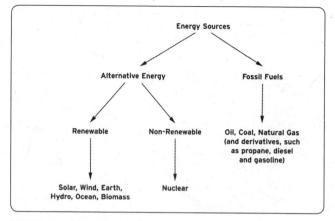

FIGURE 1.1 >>> Breakdown of Energy Sources

March of 2011, an earthquake and tidal wave devastated the Fukushima nuclear-power plant, located on Japan's northeast coast, causing radiation leaks and massive damage to its reactors. This disaster has once again raised concerns about nuclear safety and prompted many countries to reconsider their use of nuclear energy. Because it is non-renewable and presents significant risks, this controversial form of energy is not profiled in this book.

WHAT IS GREEN ENERGY?

The term *green energy* refers specifically to renewable energy sources including solar, wind, geothermal, geoexchange, hydro, wave, tidal, and biomass. Currently, fossil fuels supply the bulk of worldwide energy requirements. After fossil fuels, nuclear

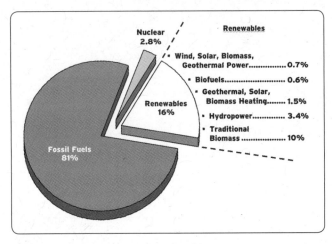

FIGURE 1.2 >>> Worldwide Energy Consumption, 2009

energy is the most consumed type of energy. As of 2009, renewable (green) energy types supplied 16 percent of global energy consumption, as shown in figure 1.2.[1]

WHY SHOULD WE CHOOSE GREEN ENERGY?

There are a number of issues driving the development of green-energy options, including economic, environmental and social factors. As the planet's growing population requires more energy and fossil-fuel sources diminish, prices for traditional energy are reaching new highs. Dependence on foreign energy supplies creates security risks such as accidental or criminal activities that could cut off those supplies. Meanwhile, human-related greenhouse-gas emissions from the extraction and combustion of fossil fuels are contributing to global warming, spurring consumers to demand cleaner alternatives to ensure a safe and healthy environment for current and future generations.

ECONOMIC REASONS

Rising Fuel Prices

The demand for energy is outpacing supply, causing prices for traditional fossil fuel–based energy to increase. Therefore, there is a pressing need to develop new sources of energy to meet this demand in order to stave off fuel shortages and resulting rising fuel prices.

The world's population is increasing and demanding more energy on a per-capita basis. From 1900 to 2000, the world's population grew fourfold while its energy consumption increased more than tenfold.[2]

Meanwhile, fossil fuel–based energy sources are declining. The world's conventional oil reserves and natural-gas reserves will eventually peak, although when this will happen is uncertain. The term "peak oil" refers to the point in time when world production reaches its highest point and then starts falling in a terminal decline. Dr. Colin J. Campbell, founder and honorary chairman of the Association for the Study of Peak Oil & Gas, maintains that the global peak for conventional oil happened in 2010, as shown in figure 1.3.[3] If we continue using the remaining oil reserves at the current rate, we could deplete them within the next few generations.

Natural gas is more abundant than conventional oil, but the reserves are distributed unevenly geographically. Most of

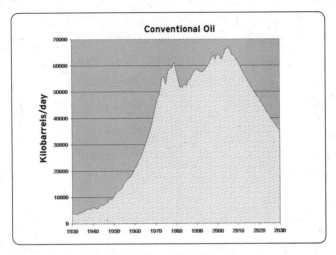

FIGURE 1.3 >>> Peak Oil Production of Conventional Oil

the major natural-gas reserves are located in the former Soviet Union and the Middle East, so greater use of natural gas by Europe and other regions could increase their dependency on the world's most politically unstable countries. Recent advances in production technology for unconventional sources, like shale gas, have made the United States the world's largest producer of natural gas, while Canada is currently ranked fifth. According to the Energy Information Administration (EIA), the United States produces enough natural gas to cover about 90 percent of its own consumption and imports about 87 percent of the balance needed from Canada.[4] Transporting natural gas is much more complex than transporting oil, since gaseous substances are more difficult to contain and move than liquids.

Coal remains plentiful but is also distributed unevenly, and its reputation as the dirtiest (most polluting) fossil fuel causes most consumers to look for alternatives that are less threatening to the environment.

Growing the Green Economy

The increasing interest in green energy is encouraging businesses to design and produce products and services for the new *green economy*—an umbrella term for a rapidly growing multi-billion-dollar sector that includes green energy, organic foods and products, green buildings and alternative-fuel vehicles. Technological innovation in green energy could play a key role in stimulating a new wave of economic growth and creating millions of new jobs.

Selling Back to the Grid

Consumers can participate in the green economy by generating renewable energy that can be sold to an energy-utility company. In some areas, individuals and small businesses are offered *feed-in tariffs* and/or *net metering* for self-generated electricity from renewable sources such as solar, wind and hydro.

A feed-in tariff is a government-imposed pricing structure that utility companies are required to pay to end users for renewable electricity that those users generate and feed into the power grid (a network of transmission lines that carry electricity to consumers). This type of tariff is designed to encourage generation and adoption of renewable energy.

Net metering, offered by some power providers, allows consumers with small renewable-energy systems to feed their unused electricity back into the grid, causing their electrical meters to run backwards. This reduces their electricity bills and can even earn them credit toward future bills. Sometimes utility companies offer net metering voluntarily, and in other cases they are required to do so by government.

Both of these incentives can stimulate regional economic development and promote growth of the renewable energy sector.

The Emerging Carbon Market

The carbon market is poised to become the next great global market. Some estimate the carbon market will grow to over $1 trillion in coming years. The *New York Times* has claimed "carbon will be the world's biggest commodity market, and it could become the world's biggest market overall."[5] Most important, the carbon market will be accessible to all, including governments, businesses and individuals.

Carbon Credits A carbon credit is a financial instrument that is transferable and saleable. One carbon credit represents one tonne (metric ton) of carbon dioxide emitted by the burning of fossil fuels.

At an international level, the Kyoto Protocol, an agreement signed by world leaders that came into effect in 2005, requires participating countries to meet greenhouse-gas emission targets and includes a provision that allows countries to buy or sell carbon credits in order to meet those targets. At regional and local levels, governments and private businesses are participating in joint renewable-energy projects and negotiating how the carbon credits earned from those emission reductions will be divvied up between the parties.

Businesses can earn carbon credits by investing in certified emission-reduction projects and then sell them to carbon-emitting organizations through exchanges such as the Montreal Climate Exchange or Chicago Climate Exchange. For example, a company could potentially earn carbon credits by building a facility that uses geoexchange heating instead of natural-gas heating and then sell the credits through an exchange to a firm seeking to reduce its carbon footprint.

Individual investors can get involved in the carbon market by investing directly in carbon credits in the hopes that those carbon credits will appreciate in value, or by investing in companies that carry out projects to earn carbon credits. For private investors, this is still a budding area of commerce; however, it's primed for growth and expansion in the future.

Carbon Offsets Unlike carbon credits, carbon offsets are not financial instruments that can be bought, sold and traded; they are more like charitable donations toward green projects. Individuals or businesses pay for carbon offsets through brokers who invest that money in projects (often involving green energy) that are certified to reduce carbon-dioxide emissions.

For example, if a person takes a vacation that involves air travel, they can go to an online carbon-offset brokerage firm that will calculate that person's share of carbon emitted during their flights and allow them to purchase carbon offsets equivalent to that amount.

Carbon Taxes vs. Cap-and-Trade Programs Various types of incentives are being used to encourage both industries and consumers to reduce their use of fossil fuels. The most common are carbon taxes and cap-and-trade programs.

There are many ways that carbon taxes can be applied. Sometimes a tax is placed on fossil-fuel energy purchased for heat, electricity and transport fuel in order to encourage both businesses and individuals to reduce their fossil-fuel consumption. Another method involves taxing utility companies on their purchases of fossil-fuel energy. Although this levy does not tax consumers directly, it is likely added to the utility bills of households or businesses. Alternatively, companies may be taxed on the amount of carbon they emit through the use of fossil-fuel energy during day-to-day operations. Carbon-tax programs are known as *revenue-neutral* if the government returns 100 percent of the revenue it collects from carbon tax to taxpayers through tax reductions in other areas.

A cap-and-trade program typically imposes a cap (in tonnes) on the level of carbon that industries or large energy users such as cement factories are allowed to emit. Companies or organizations that keep their emissions below the cap can sell their remaining allowance, converted into carbon credits, on a carbon market, while organizations that exceed the cap must buy credits or face penalties.

SOCIAL REASONS

Public Demand for Health and Safety

There is growing public demand for cleaner energy alternatives to ensure a safe and healthy environment for current and future generations. There are many diseases and health risks associated with fossil fuel–based energy production. Among these are lung irritations from nitrogen oxides, as well as lung and respiratory illnesses and cardiac problems, including arrhythmias and heart attacks, due to particulate matter such as soot and dust.

The demand for sustainability by consumers, citizens and company shareholders creates social pressure on governments and businesses to move away from fossil fuel–based energy toward green energy. Whenever there is rising social pressure for change, there also arises the opportunity for governments, organizations and individuals to take action and show leadership in their communities through clean-energy adoption and innovation.

Energy Security

Countries that import oil are vulnerable to disruptions in world oil markets, fluctuations in oil prices and political situations that could influence fuel supplies. A rise in the price of oil may cause an economic downturn for countries dependent on these supplies. If oil imports from the

Middle East to North America were held up for any reason, it could cause an oil shortage and sharp price increase that would ripple through the economy and raise the prices of all products and services dependent on that oil. This could then restrict consumers from purchasing basic products needed for their everyday lives such as food, shelter and transportation.

Both imported and domestic oil supplies are at risk of terrorist attacks and accidents that could have a profound impact on fuel supply, prices and the overall economy. Sabotage of a major domestic oil pipeline could lead to an oil shortage and a subsequent spike in fuel prices that could cause food shortages.

ENVIRONMENTAL REASONS

Oil Spills

The 2010 Gulf of Mexico oil disaster, which began with the fatal explosion and fire on the *Deepwater Horizon* drilling rig and turned into a massive oil leak, is a clear example of the environmental risks associated with traditional oil exploration. As a result of that accident, the Gulf's marine and coastal wildlife habitats, as well as the regional fishing and tourism industries, have been devastated and may not recover for decades. The risk of accident will continue and possibly increase as oil companies are forced to go into deeper water seeking oil resources that are progressively more difficult to access.

Greenhouse-Gas Emissions

Greenhouse gases (also known as GHGs) include water vapour, carbon dioxide, methane, nitrous oxide and other gases. They accumulate in the earth's atmosphere to form an invisible, blanket-like layer that traps heat within the atmosphere, similar to how glass walls trap heat in a greenhouse. An overabundance of greenhouse gases contributes to global warming. Carbon dioxide is the predominant human-related greenhouse gas in the atmosphere.

Greenhouse gases are emitted in the process of fossil-fuel combustion. Fossil fuels originate from the fossilized remains of plants and creatures that were buried in the ground hundreds of millions of years ago. Under the earth's heat and pressure, these remains decayed and transformed into hydrocarbons (a mixture of carbon and hydrogen) in the form of coal, gas and oil, which have immense energy content. Without human intervention, most fossil-fuel carbon would stay locked in the ground forever; however, when we extract and burn fossil fuels, we not only release energy for our use, we also release carbon dioxide into the atmosphere. The discovery of this energy-rich resource has led to a man-made process that unleashes massive amounts of carbon dioxide into the atmosphere every day.

Carbon dioxide serves an important role in the earth's natural carbon cycle, as it is naturally emitted by *carbon sources* such as decaying plants and animals, volcanic eruptions and animal respiration, and naturally absorbed by *carbon sinks* such as oceans and living plants. Figure 1.4 illustrates the natural carbon cycle. Living trees and other plants absorb carbon dioxide from the atmosphere during the photosynthesis process and subsequently release carbon dioxide into the atmosphere when they die and decay, or burn. Oceans, the largest natural carbon sinks on the earth, carry out complex physical and biological processes that absorb significant carbon dioxide from the atmosphere.

The carbon dioxide released by humankind's use of fossil fuels is well beyond the usual levels emitted and absorbed by the natural carbon cycle, causing overabundance of carbon dioxide in the atmosphere that contributes to global warming.

By developing alternatives to fossil fuel–based energy, we can reduce or even eliminate the use of fossil fuels, thereby

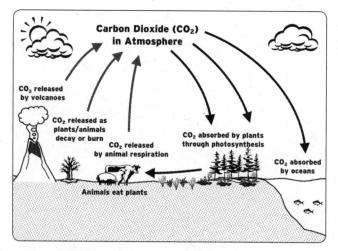

FIGURE 1.4 ⟩⟩⟩ Natural Carbon Cycle

reducing the amount of man-made carbon dioxide in the earth's atmosphere. The most effective way for society to reduce greenhouse-gas emissions would be through a massive shift away from fossil fuels to clean, renewable energy sources.

Air Pollution

In addition to greenhouse gases, fossil fuel–based energy generates other types of air pollution such as smog, ash, soot and acid rain. Although these do not contribute to global warming, they are still harmful to the environment and human health.

Smog, a brownish haze often seen over metropolitan areas, is a common form of air pollution that causes respiratory problems for humans. It is caused by ground-level ozone, fine particles and air pollution from a variety of sources including industrial emissions and vehicle exhaust.

Land and Water Pollution

Some fossil-fuel mining, transportation and production processes also wreak havoc on the environment. Coal mining causes severe environmental damage; after mining is completed, the disturbed land will remain barren unless proper topsoil is used when the area is replanted. Materials other than coal are also brought to the surface in the mining process, and these are left as solid wastes. When coal is washed, more waste material is left, and as it burns, the remaining ash is left as a waste product. Coal mining also contributes to water pollution. Coal contains pyrite, a sulphur compound; as water runs through mines, this compound forms a dilute acid that washes into nearby rivers and streams.

Production and transportation of oil also can cause water pollution. Oil spills such as the 2010 Gulf of Mexico spill often leave waterways and their shores uninhabitable for some time and result in the loss of plant and animal life.

chapter two

ENERGY 101

This chapter explains some

ENERGY BASICS such as

the forms, units and key

characteristics of energy.

FORMS OF ENERGY

Energy can be converted from one form to another, but following a fundamental law of physics known as *Conservation of Energy*, the total amount of energy in a closed system always remains the same. Therefore, energy can never be created or destroyed. Listed below are forms of energy mentioned in this book:

>>> **Radiant energy**, also known as *solar energy, solar radiation* or *sun radiation*, refers to energy coming directly from the sun.

>>> **Thermal energy**, also known as *heat energy*, refers to energy in the form of heat.

>>> **Electrical energy** refers to energy in the form of electricity, which represents the flow of electrons.

>>> **Chemical energy** refers to energy stored in molecules and released during chemical reactions such as combustion.

>>> **Mechanical energy** is the energy possessed by an object due to its motion or its position. Mechanical energy can be either *potential energy* (stored energy of position) or *kinetic energy* (energy of motion). For instance, if you hold an object above the ground, it has potential energy because of its position relative to the force of gravity pulling it downward. If the object is then dropped, the potential energy would be converted into kinetic energy as it falls to the ground.

UNITS OF ENERGY AND POWER

It's important to note that energy and power are not the same. Energy represents the amount of "work" it takes to accomplish a task, while power is the speed (rate) at which energy is generated or consumed.

ENERGY UNITS

>>> **Joule** Regardless of its form, energy is typically measured and expressed in units called *joules* (J). For example, gas bills from utility companies are calculated in joules. One gigajoule is 10^9 joules.

>>> **BTU** The *British Thermal Unit* (BTU or Btu), now replaced by the joule in most industries, represents a unit of heat equal to the amount of heat required to raise one pound of water by one degree Fahrenheit. In North America, the term is often used to describe the heating power of furnaces, stoves and barbecue grills as BTU per hour (BTU/h). 1BTU = approximately 1,055 joules.

>>> **kWh** The kilowatt hour is a unit of energy equal to 1,000 watt hours. In other words, it is the equivalent of one kilowatt of power expended for one hour. 1 kWh = 1,000 W x 1 hour x 3,600 seconds/hour = 3,600,000 J. The kilowatt hour is most commonly known as a billing unit for energy delivered to consumers by electric utility companies. MWh, GWh and TWh refer to megawatt, gigawatt and terawatt hours.

POWER UNITS

>>> **Watt** One watt of power is equal to one joule of energy consumed or generated per second. 1 W = 1 J/s

- One kilowatt is equal to 1,000 watts and one mega-watt is equal to a million watts, and so on:

- 1 kilowatt (kW) = 1,000 W = 1,000 J/s

- 1 megawatt (MW) = 1,000,000 W = 1,000,000 J/s

- 1 gigawatt (GW) = 10^9 W = 10^9 J/s

- 1 terawatt (TW) = 10^{12} W = 10^{12} J/s

>>> **BTU/h** Since power is the speed at which energy is expended, BTU per hour (BTU/h) is the power measurement associated with one BTU of energy per hour. 1 BTU/h = 0.293 W.

>>> **Horsepower** This is the conventional measurement of power, where the power of one horse equals one horsepower. 1HP = 0.746 kW.

HOW ARE GREEN-ENERGY SOURCES CONVERTED INTO HEAT AND ELECTRICITY?

There are two common methods for converting green-energy sources into electricity, space heating and/or hot water for use by consumers: by spinning turbine blades or by producing non-fossil fuels for combustion.

Electricity can be produced by using a green-energy source to create a force that spins the blades of a turbine that drives the shaft of an electric generator. For example, wood can be burned to create heat to boil water, which generates high-pressure steam that is directed into a turbine. Other examples include underwater turbines installed in tidal currents where water rushes through to spin the blades, or wind turbines erected in breezy areas where wind rotates the blades. In every case, the concept is the same—creating a force to turn the blades of a turbine that drives an electric generator to produce electricity.

Another common method for turning green energy into heat and electricity is to first produce a liquid or gaseous fuel that can be injected into an engine that drives a shaft in an electric generator to produce electricity, or that can be used directly for heating and hot water. Gaseous fuel examples include biomethane gas extracted from a landfill, biogas produced from wastewater biosolids, and syngas produced by gasifying solid wood waste, all of which can be used as direct substitutes for natural gas to provide electricity, heat and hot water. Liquid fuel examples include bioethanol and biodiesel produced from organic materials and used as a direct substitute for petroleum as a transportation fuel.

KEY CHARACTERISTICS OF ALTERNATIVE-ENERGY TYPES

When evaluating alternative-energy types, several key characteristics serve as criteria for comparison and selection:

Renewable? Can the energy source be replenished in nature and therefore continue indefinitely?

Sustainable? Does the energy type help satisfy the financial, environmental and/or social needs of the current generation without compromising those same needs of future generations?

Provides Electricity? Can the energy source be used to generate electricity?

Provides Heat? Can the energy source be used to generate heat for space heating and/or hot water?

Delivers Constant Energy? Does the energy source deliver continuous, non-fluctuating energy that's available when needed?

Widely Available Supply? Is the energy source widely available geographically?

Easily Transported and Stored Supply? Is the energy source easily transported and stored?

The following chapters examine each type of alternative energy and explain how they work, typical kinds of residential and larger-scale installations, benefits, challenges and new developments. Also, each type is evaluated based on is evaluated based on the key characteristics described above. A summary comparison of the characteristics and costs of each renewable energy type is presented in chapter twelve.

chapter three

SOLAR ENERGY

The sun's radiation is the planet's **MOST ABUNDANT RENEWABLE-ENERGY RESOURCE** and can be harnessed and transformed into electrical or heat energy for human use.

THE BASICS

The total annual solar radiation falling on the earth is more than 7,500 times the world's total annual energy consumption,[1] so even if only a tiny fraction of this energy were captured for human use, it would be more than enough to satisfy the world's energy needs.

The best places on earth to harness the sun's energy are those that receive the most direct sunlight. Excellent solar resources are available in southern Europe, North Africa and other regions of the world near the equator. Nevada, Utah, Colorado, Arizona, New Mexico, part of California and part of Texas are ideal locations in North America for using solar technology.

Capturing solar energy is an ancient practice. As early as 400 BC, the Greeks designed their houses to take advantage of solar energy. Native Americans built their homes into the sides of cliffs or hills that stored heat during the day to provide warmth at night. Also, cliff overhangs provided shade when the sun was high, and then allowed the sun to shine in during cooler times when it was low in the sky. The ancient Romans were the first to use glass windows to trap solar warmth in homes, and they also pioneered the use of glasshouses to create the right growing conditions for exotic plants.

In the 19th century, the world's first solar collector was built by Swiss scientist Horace de Saussure, and a French mathematician, Auguste Mouchout, developed a steam engine powered entirely by the sun. But interest in solar energy waned when plentiful, affordable oil and coal became available. Many years later, Albert Einstein was awarded the 1921 Nobel Prize in physics for his research on the photoelectric effect, paving the way for the modern solar-photovoltaic cell, which converts the energy of sunlight into electricity.

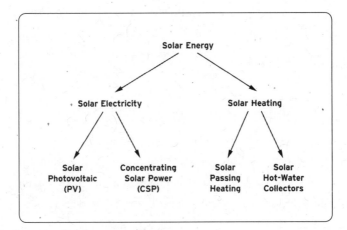

FIGURE 3.1 >>> Common Types of Solar Energy

Solar energy can be captured for either electricity or heating, and there is often confusion between the two approaches, especially since rooftop solar panels used for generating electricity look very similar to rooftop hot-water collectors used for heating. Figure 3.1 shows the breakdown of solar energy into electricity and heating, and the typical methods employed for generating each.

SOLAR ELECTRICITY

SOLAR PHOTOVOLTAIC

Electricity can be generated from solar energy in a number of ways, but the most common approach is the use of photovoltaic cells assembled in solar panels.

Solar-photovoltaic (PV) electricity is now the fastest-growing power-generation technology in the world and is being produced in over 100 countries. Global production more than doubled in 2010, reaching about 40 GW, more than seven times the capacity of five years earlier. Germany is the largest producer by far, installing more PV in 2010 than the rest of the entire world did in the previous year. Germany ended 2010 with 17.3 GW of capacity, representing about 44 percent of the world's total installed capacity. Other countries with significant amounts of installed solar-PV systems include Spain, Japan and Italy, each having about 9 percent of the world's capacity. The United States fell to fifth place with about 6 percent of the world's total capacity. PV cell manufacturing is shifting to Asia, with 10 out of 15 manufacturers now being located there.[2]

How It Works

PV cells use chemically treated silicon material that, when struck by sunlight, sets electrons free from the atoms that make up the silicon material to create an electrical current. The resulting electricity can be used immediately or stored in a battery until needed at a later time. The electricity produced from PV technology is direct current (DC), so must be converted to alternating current (AC) with an inverter in order to work with household appliances.

The energy efficiency of solar-PV panels currently available on the market is about 15 to 20 percent, which means they convert only 15 to 20 percent of the solar energy striking the panels into electrical energy. The efficiency is mainly limited by present silicon-panel technology. Other conditions that can limit efficiency include the angle of the solar panel in relation to the sun's rays, shading (from trees, adjacent buildings, etc.) and cloud cover. To maximize the amount of sunlight striking a panel, it is important to install it at right angles (90 degrees) to the sun's rays in a location that receives lots of direct sunlight.

Real-Life Applications

Residential solar-PV systems can be either off-grid or connected to a power grid. Solar electricity is typically generated on a fluctuating basis due to intermittent sunshine; therefore, in order to maintain an available supply of electricity at all times, PV systems must have access to a power grid or have batteries to store the electrical energy supplied by the system.

Off-grid solar-PV systems are often used for homes in remote areas. In some cases, they use DC generated by solar panels to operate electrical appliances that have been especially designed for DC; however, off-grid systems more typically use batteries to store electrical energy and power an inverter that converts DC to AC for operating standard appliances.

Residential solar-PV systems connected to a power grid are able to purchase electricity from the grid whenever they have a shortfall, which could occur during periods of high demand or low sunlight. If the utility company offers net metering, the consumer's electrical meter will run in reverse when they are generating excess electricity and feeding it back into the grid, reducing their electricity bills. If the consumer is actually putting more electricity into the grid than they are taking out, they may be allowed to use that surplus to build up credit toward future utility bills or in some cases even sell the surplus in return for cash payments. PV systems connected to a power grid require that the DC be converted into AC before it is fed into the grid. Figure 3.2 illustrates a typical residential solar-PV system connected to a grid.

For commercial applications, solar-PV power plants typically use a large, linked collection of PV panels known as an *array*. Fedex has installed an extensive array of PV panels on the roof of its Oakland, California, operation. This system has

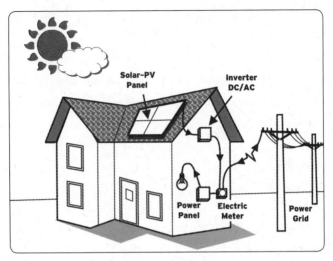

FIGURE 3.2 >>> A Typical Residential Solar-PV System

a capacity of more than 900 kW, the equivalent of more than 450 home-rooftop systems.[3]

In addition to power plants, other commercial applications for PV technology include municipal street lights, parking meters and maritime buoys, to name a few.

Benefits

Solar-PV panels are noise-free, have no moving parts and generate no greenhouse gases, toxic compounds or other pollution. They produce maximum power exactly when it is needed—during the day when electricity demand and prices are the highest. Furthermore, rooftop solar-PV panels are inconspicuous compared to other types of energy systems.

Best of all, solar energy is available everywhere, reducing or even eliminating the need for transmission lines. This is significant because power lines are not only costly to construct, but the construction process damages the environment, and the resulting power lines may even pose safety risks to nearby residents.

Challenges

With all the advantages of solar PV, you may wonder why PV panels aren't on every roof. The main obstacles to wider implementation of solar-PV technology are relatively low energy efficiency (typically 15 to 20 percent) and high manufacturing costs that combine to make it a relatively expensive source of electricity. Because of their low energy efficiency, multiple PV panels are often required to generate enough electricity to meet the demands for a building. The number of panels needed depends on the building's location and how much power it consumes. In some cases, it may be necessary to cover an entire roof with solar-PV panels to generate enough electricity to meet the needs of a household or business.

Of course, weather plays a role in the effectiveness of a solar-PV system. Regions closer to the equator have more hours of sunlight per day, so deliver better results. Cloudy and smoggy areas still have lots of solar energy, but it is diffused enough to make a difference in the energy output of a PV system.

The financial feasibility of solar-PV electricity also depends on the local price of the electricity it displaces. In regions where electricity is expensive, such as California, solar PV is more likely to be considered a feasible option. If you live in a place where inexpensive electricity is available from the local grid, there will likely be fewer solar-PV installations, and homeowners who do invest in rooftop solar-PV panels may not attain a full return on their investment while living in those homes.

During a press conference at the 2010 Olympic Games in Vancouver, British Columbia, a journalist asked, "We heard this was one of the greenest cities, and it's supposed to be hosting the greenest Olympics, so why don't we see solar-PV panels everywhere?" Although Vancouver implemented many green initiatives for the Olympics, it did not use much solar-PV technology because Vancouver has access to very inexpensive hydropower and has limited sunshine. This means that solar-PV systems are less feasible in Vancouver than in cities with more sunny days and higher electricity prices.

New Developments

The holy grail of solar-PV technology is a solution that increases energy efficiency while reducing production costs, and many research and development programs around the world are working toward that goal.

Research is being conducted on new semiconductors that do not use silicon, one of the most expensive elements of a PV system. Also, research is being conducted on PV multi-spectral layers, which can absorb more solar radiation, thus increasing energy efficiency, although they come at an increased cost.

Rooftop solar-PV panels should be installed where they receive maximum exposure to the sun. Elena Elisseeva/iStockphoto.com

There are several competing types of photovoltaic cells. While crystalline-silicon cells (known as first-generation PV cells) are most widely used today, there is growing use of thin-film PV-cell technology (known as second-generation) for rooftop solar panels and other applications. Thin film is a slim, lightweight and flexible PV-cell material that can be used as a surface covering, thereby serving two purposes: roof/wall covering and power generation. Thin-film technology is relatively inexpensive; however, it currently has lower energy efficiency than crystalline silicon–cell technology. Development is under way to increase the efficiency of thin-film technology.

Some believe third-generation PV-cell technology may someday allow for PV cells to be painted directly onto car surfaces to power vehicles or onto building surfaces to power their operations.

OTHER METHODS OF GENERATING SOLAR ELECTRICITY

Concentrating solar power (CSP) is another solar-power technique that uses mirrors or lenses to concentrate sunlight onto tubes or tanks of water or some other fluid until it boils, creating high-pressure steam that turns the blades of a turbine, which drives an electric generator to produce electricity. A common application of CSP is a parabolic trough, which acts as a reflector that concentrates light onto a focal point where a tube filled with fluid is located. CSP is currently used in large-scale power generation systems rather than residential applications. In 2010, 478 MW of new CSP capacity was added, bringing the total global capacity to 1,095 MW. Spain took the lead by adding 400 MW to bring the country's total capacity to 632 MW. As of April 2011, another 946 MW were under construction in Spain. The United States fell from first to second place, ending the year with 509 MW after adding only 78 MW in 2010; however, as of early 2011, a further 1.5 GW of CSP capacity was under construction in the United States, and contracts have been signed for another 6.2 GW of capacity as a result of federal incentives and mandates.[4] The largest operating solar energy–generating facility in the world, the Solar Energy Generating Systems (SEGS) facility located in California's Mojave Desert, uses parabolic-trough technology, whereby sunlight bounces off mirrors and is directed to a central tube filled with synthetic oil, which heats to over 400°C. The synthetic oil transfers its heat to water, which boils and drives a steam turbine to generate electricity. Synthetic oil (instead of water) is used to keep the pressure in the system manageable.

Concentrated photovoltaics (CPV) systems, which employ even newer technology than PV and CSP systems, use optics such as dish reflectors to concentrate a large amount of sunlight onto a small area of solar photovoltaic material to generate electricity. In comparison to conventional flat-panel photovoltaic systems, CPV systems require a much smaller area of solar cells. Developers of these systems claim that they combine high-efficiency solar cells with advanced concentrating optics to provide high energy yield using just 1/1000 of the amount of solar-cell material used in traditional photovoltaic systems.

SOLAR HEATING

Solar heating captures the sun's radiant energy and converts it into thermal (heat) energy for either domestic hot-water use and/or space heating. There are two ways of capturing solar heat: passive and active.

PASSIVE SOLAR HEATING

Passive solar heating refers to solar heating achieved without the use of powered devices, such as pumps and fans, by orienting a building so its windows allow the maximum amount of solar energy to naturally heat its interior. The key objectives for passive solar design are to allow sunlight into interiors when light and warmth are needed and to keep sunlight out during hot weather.

The techniques used to meet these objectives depend on whether a building is located in a warm or cold climate, but effective passive solar heating is quite easily achievable without dramatic cost increases when building a new home. Here are some general guidelines:

>>> Design the building to be longest along the east/west axis.

>>> Orient the building to face south or a few degrees to the east to capture the morning sun.

>>> Place trees strategically to provide shade in the summer, especially along the west side.

>>> Carefully plan the placement of internal walls, rooms and doors to allow plenty of natural light into main living areas.

>>> Install most windows on the south side; use large windows that let in the winter sun but are shaded from the high summer sun.

>>> Use appropriate window glazing to control solar heat gain.

>>> Use roof overhangs and other shading devices (trellises, shutters, etc.) to reduce summer heat gain and ensure winter heat gain. For example, a strategically positioned roof overhang provides shade for south-facing windows in the summer when the sun is high in the sky, but allows sunshine to reach the windows in the winter when the sun is low in the sky.

>>> Install the appropriate amount and type of insulation to minimize seasonal excessive heat gain or loss.

>>> Use a south-facing masonry wall or other structure that absorbs and stores solar heat and then radiates it into the interior living space during late afternoon and evening.

SOLAR HOT-WATER COLLECTORS

In contrast to passive solar-heating systems, active systems use powered equipment to capture, transfer and store heat. Solar hot-water collectors are common active solar-heating systems. They are installed on south-facing roofs, allowing water to be heated by the sun as it circulates through them.

How It Works

Solar collectors vary in design, but one common type is known as a *flat-panel collector*. It is typically composed of an insulated black box with a flat glass top that contains a network of piping. Water is heated by the sun as it passes through the piping in the box, thus converting solar energy to thermal energy. The piping system is connected to a water-storage tank and a pump, allowing water in the piping to circulate between the collector and the tank, where hot water is stored until needed. Figure 3.3 shows the configuration for a typical solar flat-panel collector system.

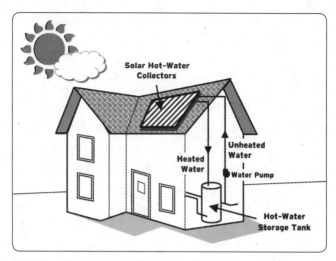

FIGURE 3.3 >>> A Typical Solar Hot-Water Flat-Panel Collector System

Real-Life Applications

Solar hot-water collector systems are often used for domestic hot-water purposes such as showers, washing machines and dishwashers. These systems are typically sized to provide 100 percent of hot-water needs in the warmer months and are supplemented by boilers or gas-fired hot-water tanks in colder months. In addition to providing hot water, a collector system can also provide space heating, typically achieved by distributing the hot water throughout a building through wall-mounted radiators or radiant floors. In some cases, a heat exchanger is used to transfer heat from the hot water to air, which is then delivered throughout the building via air ducts to provide space heating.

Unlike residential solar-heating systems, commercial applications usually have a much larger, steadier demand that makes solar heating an even more feasible investment. Today, solar hot-water heating is being applied to hundreds of commercial applications that require "process water heating," a term used to describe industrial water-heating applications. The principles of industrial solar hot-water heating are the same as for residential applications, except they generally have more collectors, bigger pumps and larger storage tanks. Many commercial facilities have flat roofs that can be used for a large number of solar collectors. Examples of commercial applications for solar hot water include car washes, laundromats, hotels, restaurants, breweries, apartment buildings, indoor pools, food-processing facilities and agricultural sterilization operations. For example, Chanterelle Inn, a large country inn located in Nova Scotia, uses a solar hot-water system consisting of 16 hot-water collectors, plus two solar-PV panels to power the system's water pump. Operating year-round, the two-storey inn features eight suites on the upper floor, as well as a kitchen, dining room, lounge and another large suite on the main floor. Although there is an electricity-based heating backup system available if needed, the building consumes no fossil fuels on site.[5]

Benefits

Solar hot-water collectors are arguably the most cost-effective and easy-to-install green-energy systems, especially for residential purposes. Solar hot-water systems typically achieve 50 to 80 percent efficiency, which means they convert 50 to 80 percent of the radiant energy striking the collectors into usable heat energy. Solar collectors are relatively inexpensive to purchase and install compared to other types of renewable energy. With its relatively low cost and high efficiency, solar-heating technology typically offers a much better return on investment than solar-PV technology and other green-energy options. Therefore, it can be feasible even in locations that receive limited sunshine.

Solar hot-water collectors have a long, proven track record and have been widely used in many countries for years. In fact, solar hot-water systems are mandatory in all new construction in Israel. Spain mandates that 60 percent of hot-water heating must come from solar systems in all new buildings and major renovations in Barcelona. Brazil requires all public buildings in Rio de Janeiro to use solar hot water for at least 40 percent of heating energy. Japan requires property developers to assess and consider possibilities for solar hot water and other renewables and to report assessments to owners in Tokyo. China requires solar hot water in all new residential buildings up to 12 stories.[6] In the United States, Hawaii became the first state to require solar hot-water collectors on new homes starting in 2010. Hawaii currently relies on imported fossil fuels more than any other state, with about 90 percent of its energy sources coming from foreign countries.

Challenges

The solar hot-water heating systems described above require sunlight, so weather plays a role in their level of energy output. Like solar-PV systems, solar hot-water heating is less efficient in cloudy and smoggy regions, where solar radiation is diffused.

Collectors can be damaged by bad weather, such as hail, and by falling debris, such as tree branches. They must also be kept clear of snow buildup, which may require routine trips up to the rooftop.

New Developments

As conventional energy prices rise, there is growing interest in solar heating, and new systems are being developed that are easier to install and less expensive to operate. Solar-heating systems are now being used more in multi-storey construction and are being combined with PV systems to produce both heat and electricity from the same surfaces.

Hybrid photovoltaic-solar thermal collectors, sometimes known as *hybrid PV-T systems,* produce electricity and hot water simultaneously. These systems combine a PV cell, which converts solar radiation into electricity, with a solar-thermal collector, which captures the remaining heat for space heating and hot water. For example, the Sonoma Wine Company in northern California uses a hybrid PV-T energy system for its operation, which processes more than 7,500 tonnes of grapes, stores and services 65,000 barrels of wine, and bottles 4 million cases each year. The hybrid system was installed at the company's site by a solar-system developer who continues to own and operate the system while Sonoma purchases the heat and electricity at guaranteed rates. This solar cogeneration system displaces about 64 MWh of electricity and about 1,320 gigajoules of natural gas annually to heat water for wine processing, sanitation and barrel-services operations.[7]

OTHER METHODS OF GENERATING SOLAR HEATING

Evacuated-tube collectors are an alternative to flat-panel hot-water collectors. They contain rows of transparent glass tubes with the air pumped out of them (evacuated) to create a vacuum, which reduces heat loss, thereby making it possible to achieve higher temperatures. Inside each glass tube is a flat metal plate (absorber) fused to a heat pipe that extends out one end of the tube into a header pipe filled with flowing water or fluid. Heat from the sun is absorbed by the absorber and transferred to the heat pipe by conduction. Then the heat is transferred from the heat pipe to the fluid inside the header pipe by convection. The fluid is then circulated through a domestic hot-water or radiant space-heating system.

Generally, flat-panel collectors are less expensive and more efficient at low-to-moderate temperature differentials (the difference between internal water temperature and outside air temperature), while evacuated-tube collectors are more efficient at higher temperature differentials. Since most residential solar-heating applications do not require very high temperatures, they tend to be well suited to flat-panel collectors. However, evacuated-tube collectors typically perform better for commercial buildings that require very high temperatures for certain applications such as sanitation purposes.

Another method of generating space heating from solar energy is with a *solar hot-air collector* (instead of a hot-water collector) that circulates building air through a solar air-collection box where the air is heated and then directed back into the building. One example of a solar hot-air collector system is a solar wall unit installed on an external south-facing wall of a building where it absorbs heat from direct sunlight to warm the air inside the unit. That air is then circulated through the building, using electric fans, to provide space heating. This type of system is relatively inexpensive to purchase and install, but because it does not store heat energy, the heat is only available on sunny days. It is not available on overcast days or at night, when it is more likely to be required. Since this kind of system does not always provide heat when it is needed, it must be supplemented. However, this type of solar wall unit can effectively provide heating on cool, sunny days as just one part of a larger, comprehensive energy system.

Renewable? Yes, solar energy is renewable because there is an endless supply of sunshine.

Sustainable? Yes, solar energy is sustainable since it can deliver energy to the current generation without compromising future generations.

Provides Electricity? Yes, solar energy is used to generate electricity, usually through the use of PV technology, or by generating heat that boils water to drive a steam generator.

Provides Heat? Yes, solar energy is also used to generate heat and/or hot water, typically by using rooftop hot-water collectors or solar air-heating systems.

Delivers Constant Energy? No, the sun does not shine continuously, so solar systems deliver fluctuating energy.

Widely Available Supply? Yes, sunshine is available in all regions around the world, although some regions receive more than others.

Easily Transported and Stored Supply? No, sunshine cannot be transported or stored. However, after solar energy has been converted to electricity, then the electricity can be fed into transmission lines and transported to where it is needed. Hot water can only be moved short distances and stored for a brief time before losing its heat.

chapter four

WIND ENERGY

Wind contains **"ENERGY OF MOTION,"** known as kinetic energy, which can be captured and transformed into **ELECTRICAL ENERGY** for use in homes and industries.

THE BASICS

Wind is caused by differences in air pressure due to the uneven heating of the earth's surface by the sun. The earth's surface absorbs sunlight and radiates it back out in the form of heat. Some parts of the earth absorb more solar energy than others, depending on various factors such as the angle of the sun's rays and the type of surface. Land surfaces usually absorb more solar energy than bodies of water, so the air over land usually gets warmer than air over water. As air warms, it expands and rises, and nearby cooler, denser air rushes in to take its place, creating wind. In the same way, the large atmospheric winds that circle the earth are produced because the equator receives more direct sunlight than the poles, so the surface air over the equator is warmer than that over polar regions.

Wind could supply the world's electric power requirements 35 times over and all energy requirements 5 times over.[1] Even if only about 20 percent of this power could be captured, it would satisfy 100 percent of the world's energy demand for all purposes. In other words, global wind resources are much greater than the total energy needed by humanity at this time.

There are large untapped wind resources on every continent, especially on hilltops, open plains, mountain passes and near the coasts of oceans or large lakes.

According to the World Energy Council, global offshore wind resources are also significant, with European offshore wind potential, for example, capable of supplying all the European Union's electricity needs without going farther than 30 kilometres offshore. Typically, offshore winds are faster and more consistent than onshore winds at lower heights due to the reduced surface roughness over the ocean.[2]

Global wind-power capacity (generated from installed wind-turbine systems) reached about 198 GW after growing by 39 GW in 2010, more than any other type of green energy. China surpassed the United States as the wind-power leader by adding 18.9 GW, which accounted for 50 percent of new global capacity in 2010. Germany, the wind-power leader in Europe, reached 27.2 GW in 2010.[3] Canada reached about 4.6 GW,[4] enough to power over a million Canadian homes, after strong growth in 2009 and 2010 that culminated in Canada now generating wind electricity in every province. The United States and Canada together account for about 15 percent of the global wind-power market. Figure 4.1 also shows the installed wind-power capacity of other top-ranking countries at the end of 2010.[5]

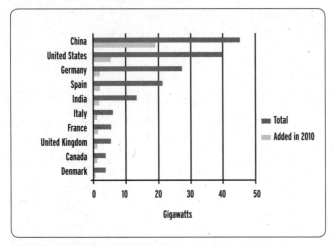

FIGURE 4.1 ≫ Installed Wind-Power Capacity: Top 10 Countries in 2010

WIND TURBINES

Humans have been using wind turbines to harness the energy of wind for centuries. The first wind turbines were used for mechanical instead of electrical energy, since their blades would spin to drive the shaft of a mechanical device such as a grain grinder or water pump instead of an electric generator. The first known windmills, in Sistan in eastern Persia, date back to the 7th century. Windmill use spread across the Middle East and Central Asia, and later to China and India. By 1000 AD, windmills were used to pump seawater for salt-making in China and Sicily, and by the 1100s, they were used extensively in northwestern Europe to grind flour. By the 14th century,

Dutch windmills were used to drain areas of the Rhine River delta. Thousands of windmills were used in Denmark by 1900 to provide mechanical energy for pumps and mills, and a large number of small windmills were also installed on North American farms to operate irrigation pumps.

Windmills were first used to produce electricity around 1887, when Professor James Blyth of Anderson's College in Scotland designed and built a cloth-sailed wind turbine to provide power to his own cottage. Meanwhile, across the Atlantic, Charles Brush was building a larger wind-power turbine at his home in Cleveland, Ohio. By the 1930s, windmills could typically be found producing electricity on American farms. Utility-scale wind-turbine power systems emerged in the mid-20th century.

How It Works

The basic components of a wind turbine include a tower, blades (rotors), shafts, gearbox and generator, as shown in figure 4.2. As wind passes over the turbine blades, it exerts a force that rotates the blades, which turn a shaft connected to a gearbox. The gearbox increases the shaft-rotation speed to a level suitable for an electric generator, which then converts the mechanical energy in the rotating shaft to electrical energy. The resulting electricity flows down the tower and into a transformer that changes it to the required voltage before it's fed into the grid.

Winds are stronger and more constant farther from the earth's surface, so wind turbines are positioned on tall towers.

The electricity generated by wind turbines is often collected and fed into power lines (the power grid), where it is combined with electricity from other sources and then delivered to consumers.

Electricity output from a wind turbine is strongly dependent on the blade size. Wind power captured by a turbine is proportional to the square of its blade length. For example, doubling the blade length increases the power output by four.

As you might expect, wind speed is another key factor in turbine performance. Typically, small wind turbines require an annual average wind speed greater than 14 kilometres per hour, while utility-scale wind-power farms require steady winds that average at least 20 kilometres per hour. The power available in the wind is proportional to the cube (x^3) of its speed; therefore, a small difference in wind speed results in a large difference in the amount of electrical energy generated.

A 10-kW wind turbine can generate about 10,000 kWh annually at a site with average wind speeds, which is approximately enough to to power a typical household (excluding air conditioning). Most modern, large-scale

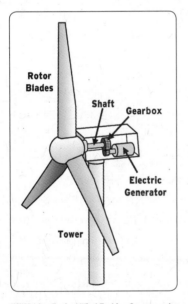

Rotor Blades

Shaft

Gearbox

Electric Generator

Tower

FIGURE 4.2 ⟫ Typical Wind-Turbine Components

installed wind turbines have power ratings ranging up to 1.8 MW and can produce about 5.2 million kWh in a year, enough to power approximately 500 households.

Real-Life Applications

A small-scale turbine system connected to a grid can power a house when it's windy. When the wind isn't blowing, the house will use electricity supplied by the local utility company through a typical grid connection. If the utility company allows net metering, any excess electricity generated from the wind-power system will be fed into the grid. Figure 4.3 shows the typical configuration of a residential house using a wind turbine connected to a power grid.

Smaller wind turbines are often installed at cottages or rural homes in remote areas without access to a power grid. Unlike solar-PV systems, wind turbines generate AC power, which can be used directly with most home appliances. Any excess energy is usually stored in rechargeable batteries, which requires the energy to be temporarily converted into DC power. When needed, the electricity is fed from the batteries into an inverter that changes the output from DC back to AC. During times when the batteries have been completely discharged and there is not enough wind to generate electricity, a backup system would be required to avoid a blackout.

Wind farms range from medium-sized operations, commonly found in farming communities, to massive operations that generate hundreds of megawatts of electricity,

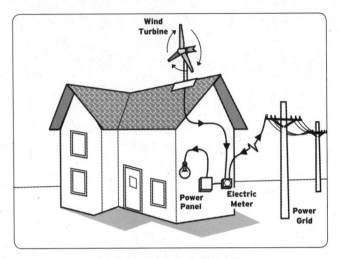

FIGURE 4.3 >>> A Residential Wind Turbine Connected to a Grid

typically located in remote mountain ranges or on large, open plains.

An example of a commercial wind-power operation found in a farming community is the Fenner Wind Farm in Madison County, one of the first wind farms in New York State. It operates 20 turbines to harness wind power and supply renewable energy across the state. The farm consists of 1.5-MW turbines generating a total of 30 MW of power.[6] Farmers there say that it's just like any other crop, except it's much less work to harvest and presents lower financial risk: "As long as the blades are turning, we know that we'll get paid."[7]

In the United States, the three states with the most installed wind capacity are Texas, Iowa and California. The world's largest wind farm, Roscoe Wind Farm, is located in Texas and has over 600 turbines producing more than 700 MW of electricity, enough to power about 250,000 homes.

In Canada, there are currently just over 100 wind farms with a total combined capacity of more than 3,400 MW, enough to power over 1 million homes. Most of these wind farms are located in Alberta and Ontario; the largest is the Melancthon EcoPower Centre near Shelburne, Ontario, which provides 199.5 MW of power. In Canada, wind farms now produce enough power to meet roughly 1 percent of Canada's total electricity demand.[8]

Currently the largest offshore wind farms in the world are located in the United Kingdom and Denmark, with the biggest one being the Thanet wind farm, 11 kilometres off the coast of the Thanet district in Kent, England. Many new offshore wind farms are under construction, including the 500 MW Greater Gabbard wind farm off the Sussex coast in the United Kingdom and the 400 MW BARD Offshore 1 wind farm in Germany.

Benefits

Wind energy has been used for centuries and is one of the most promising sources of electricity for the coming years. It is widely available in most countries and does not have to be purchased like many traditional energy resources such as oil. Furthermore, wind farms can be deployed in a matter of months, assuming an available supply of turbines, and can start generating power and income as soon as the first turbine is connected to the grid. Unlike power plants that take years to build and require large capital investments, wind farms can be implemented incrementally; farms can start with a small capital investment of just a few turbines and then grow over time by simply adding more turbines.

Turbine component costs have come down dramatically over the last 15 years due to increasing production volumes. Also, the efficiency and reliability of modern wind-turbine technology has significantly improved, bringing the overall cost of wind energy closer to the cost of conventional energy sources.

Wind turbines do not emit greenhouse gases or other air pollutants when they are operating. Although there are some greenhouse-gas emissions during the construction of turbine components, those emissions are easily offset by the system's many years of carbon-free operation once it is up and running.

Wind turbines are now becoming a common sight in the Alberta foothills.
wildroze/iStockphoto.com

Challenges

Like any source of electricity, wind energy has some drawbacks. It is currently still more expensive, on a per-kilowatt-hour basis, to generate electricity from a wind-power facility than a traditional fossil fuel–powered facility. This makes it challenging to run a profitable wind farm in some locations. One reason for the higher cost is that the level of production for wind-turbine components has not reached sufficient volume to achieve economies of scale. Also, the tall turbine towers and large blades are difficult and costly to transport, and installation requires expensive cranes, skilled operators and experienced installers. It is often expensive to build the infrastructure to transmit the electricity from remotely located wind turbines to utility power lines, and this construction, including roads and transmission lines, may have a significant impact on the environment.

The commercial viability of wind power also depends on the existing electricity prices in a region. Because of large hydro dams, many Canadians enjoy some of the lowest electricity prices in the world, so wind farms may find it difficult to compete with those prices, even with subsidies. Electricity prices in the United States and elsewhere vary greatly from region to region, with some areas having fossil fuel–based electricity prices that are similar to wind-generated electricity prices.

Weather and wind variations make it difficult to predict monthly wind capacity and associated power output. These fluctuations are not popular with utility companies, which prefer predictable, steady power availability. In fact, some utility companies apply financial penalties to wind-power producers when they don't deliver a predetermined level of monthly power, making the wind-power business a risky investment for those facing such penalties.

Wind turbines are highly visible on the landscape. Some people think they spoil the view in otherwise scenic areas, while others appreciate their majestic stature and perceive them as elegant symbols of a better, greener planet. There have also been complaints about turbine noise. Again, this is a subject of debate. Farmers who live near the Fenner Wind Farm say the turbines are quieter than the sound of a car driving by. Any noise problem could be averted by locating wind turbines some distance from buildings, as is done with other power-generating facilities.

Another common concern about wind farms is that birds and bats are sometimes killed or injured as they fly into or near the turning blades. A University of Calgary study determined that most dead bats found under wind turbines near Pincher Creek, Alberta, died from barotrauma—physical injuries caused by a sudden drop in air pressure—that occurs when they fly too close to wind turbines. Wind turbines are now being designed with larger, slower-rotating blades that are easier to see and avoid, and the turbines' towers are being positioned well away from known bird and bat migration flight paths. Also, because bats are more active when wind speeds are low, one proposed strategy is to increase the wind speed at which turbine blades begin to rotate during the bats' migration period.

A 2009 University of Singapore study concluded that, on a worldwide basis, fossil-fuelled power stations appear to pose a much greater threat to birds than wind turbines.[9] Proponents of wind energy say that birds (and humans) face much more serious life-threatening consequences from climate change than from wind farms, which help reduce climate change. Furthermore, according to the UK's Royal Society for the Protection of Birds, "If wind farms are located away from major migration routes and important feeding, breeding and roosting areas of those bird species known or suspected to be at risk, there is a strong possibility that they will have minimal impact on wildlife."[10]

Offshore wind farms are under scrutiny for how their infrastructure affects marine life and ecosystems. Their proponents are quick to respond by pointing out the far worse effects of offshore oil rigs, as witnessed during the catastrophic 2010 Gulf of Mexico oil spill. Various regulations covering habitat and wildlife conservation are in place to ensure minimal environmental impact from the construction and operation of offshore wind farms. However, building permits for offshore wind turbines depend on a large number of agencies and institutions such as navigation, national parks, pipelines, cables, defence areas, etc. Many European countries have appointed one authority to coordinate the various organizations involved. The Canadian government has not developed regulations specific to offshore wind farms but uses existing statutes to regulate their various aspects. In the United States, the Minerals Management Service has lead jurisdiction over permitting, competitive leasing and royalty payments for offshore wind development on the outer continental shelf.

New Developments

The newest, largest turbines under construction have a capacity of 5 to 7 MW, enough to support up to 1,700 households in North America. They have computerized guidance systems that turn the direction of the turbines and change the pitch of the blades to maximize electrical output. Researchers are also exploring designs to improve the structural dynamics and aerodynamics of wind-turbine blades in order to increase energy yield and thereby reduce the cost of a wind turbine per unit of energy. Improvements in the design and efficiency of other turbine components such as gearboxes are also ongoing.

The cost of turbine materials is slowly coming down as the technology evolves. There is a trend toward lighter-weight, low-cost materials for the tower, since it represents the heaviest component of the turbine. Another technology shift is occurring in the drive train; in some cases, the gearbox is being completely eliminated by using variable-speed generators and solid-state electronic converters that produce AC power suitable for power grids. This trend began in smaller turbines and is now being incorporated into larger ones.

Vertical-axis wind turbines (VAWTs), which have blades that rotate around a vertical axis, offer an alternative to the traditional horizontal-axis wind turbines. The main advantage of VAWTs

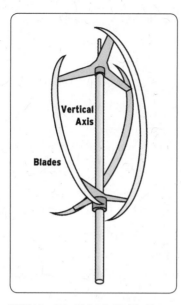

FIGURE 4.4 ›› A Vertical-Axis Wind Turbine

is that they do not need to be directed into the wind to be effective, so they are ideal at sites where wind direction is highly changeable. Figure 4.4 illustrates a typical vertical-axis wind turbine.

VAWTs are usually mounted closer to the ground than horizontal-axis wind turbines, which are on tall towers; therefore, the generators and gearboxes are more accessible for maintenance. In most cases, VAWTs require less speed to start up and generate less noise. However, they do not produce as much energy at a given site as a traditional horizontal-axis wind turbine with the same footprint because they are mounted closer to the ground where less wind speed is available, and they have inefficiency from dragging each blade back through the wind. VAWTs are typically used for residential or small commercial applications in locations where wind speed is low or wind direction changes frequently. For example, in 2010, Edgecombe Community College in North Carolina installed a VAWT on the campus greenway as part of the college's Earth Day celebration. "This looks nothing like a traditional windmill," says George Anderson, a sustainability coordinator at the college. "In fact, it looks more like a piece of art than a structure that generates energy."[11]

A new concept designed to smooth out the variability of wind energy combines a wind-turbine system with a hydro-energy storage system to deliver steady power to the grid. The approach uses excess wind power during windy periods to pump water to an elevated, dammed reservoir and later releases that water through hydro turbines to generate hydropower when the wind is not blowing.

Finding other ways to store fluctuating wind energy and deliver steady wind power would have a significant impact on the wind-power industry. A New Jersey electricity company is partnering with inventor Michael Nakhamkin to develop a way to store wind-generated power in underground reservoirs. If successful, the technology would store excess wind power underground, in the form of compressed air, and later release it when the wind isn't blowing.

KEY CHARACTERISTICS OF WIND ENERGY

Renewable? Yes, wind will continually be produced in nature as long as the sun shines on the earth.

Sustainable? Yes, wind is sustainable, since the wind energy we capture for use by the current generation does not have a negative effect on future generations.

Provides Electricity?   Yes, wind generates electricity by rotating turbine blades, which turn the shaft of an electric generator.

Provides Heat?   Not directly, although the electricity generated by a wind turbine could be used to power an electric heater.

Delivers Constant Energy?   No, wind fluctuates.

Widely Available Supply?   Yes, wind is available in most regions of the world.

Easily Transported and Stored Supply?   No, wind itself cannot be transported or stored. After wind energy been converted to electricity, the electricity can be fed into transmission lines and transported to where it is needed.

EARTH ENERGY

Energy can be extracted

from **BENEATH THE**

EARTH'S SURFACE

to provide electricity

and/or heat for

human use.

THE BASICS

Earth energy can be divided into two types, geothermal and geoexchange (also known as ground-source), which both derive their energy from underground. The main difference between them is that geothermal energy originates from thermal energy at the earth's core, while geoexchange energy originates from

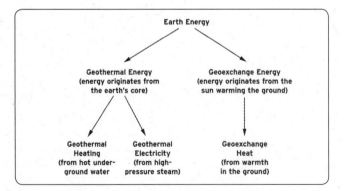

FIGURE 5.1 >>> Types of Earth Energy

solar radiation absorbed within the ground. Furthermore, geo-thermal energy is used mainly for generating electricity as well as providing heat, while geoexchange systems are used to provide heating and cooling only. Figure 5.1 illustrates these two types of earth energy.

GEOTHERMAL ENERGY

Geothermal literally means "heat from the earth," and refers to heat energy that originates in the earth's core. In certain locations, hot water or high-pressure steam exists in the earth's crust close enough to the surface to be accessed and used to provide electricity and/or heat for human use.

Temperatures at the centre of the earth are estimated to reach about 5,600 Kelvin[1] (about 5,327°C), high enough to melt solid rock into magma. Magma is less dense than rock, so it rises to the surface. Occasionally, magma escapes through volcanic eruptions, but usually it stays beneath the earth's surface, heating surrounding rocks and the water trapped between those rocks. Sometimes that water escapes through cracks in the earth to form pools of hot water (hot springs) or gushing hot water and steam (geysers). The remaining heated water sits under the earth's surface in pools called geothermal reservoirs.

The amount of thermal energy in the earth's core is enor-mous, so it is sure to play an important role in the future energy mix for certain countries. Geothermal resources are usually found where the earth's tectonic plates interact and geologic conditions allow large amounts of magma to rise to the earth's surface, such as the "Ring of Fire" circling the Pacific Ocean. Some countries along this ring, including Indonesia, the Philippines, Japan, the United States and Mexico, are already using their geothermal resources for electricity production. Since the required conditions do not exist everywhere, there are limited opportunities to use geothermal power on a worldwide basis.

The first geothermal power plant was built at Tuscany, Italy, in 1904 at a place where steam was spewing out of the ground. Although the plant was destroyed in the Second World War, it has since been rebuilt and expanded and is still producing electricity today.

Currently, there are over 11 GW of installed geothermal power capacity in at least 24 countries,[2] and as many as 70 countries have projects either under development or under active

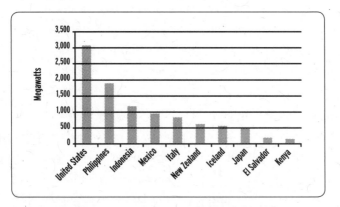

FIGURE 5.2 >>> Top 10 Countries Generating Geothermal Power in 2010

consideration.[3] The top 10 countries with existing installed capacity in 2010 are shown in figure 5.2.

The United States is the world's largest producer of geothermal power, with over 3 GW of capacity from more than 70 plants in 9 states. As of 2010, the Philippines was the world's second-largest producer, with over 1.9 GW of installed capacity, representing approximately 18 percent of the electricity used by that country. Up to 40 percent of the world's geothermal potential is located in Indonesia, which had over 1.1 GW of installed geothermal power capacity in 2010, with a goal of 9.5 GW. Iceland derives 25 percent of its electricity and 90 percent of its heating needs from geothermal resources and in 2010 had over 500 MW of installed geothermal power capacity. The country benefits from the relatively close proximity of near-surface geothermal reservoirs to urban communities and is considered a model for geothermal development.[4]

Canada has not yet developed any of its geothermal energy resources; however, there are a number of geothermal projects in development in British Columbia and Alberta. The country's most advanced project, located about 170 kilometres north of Vancouver, British Columbia, is known as South Meager and is expected to provide about 100 MW of geothermal power when fully operational in 2012.

How It Works

Geothermal steam is captured and directed through the blades of a turbine, which rotates a shaft connected to an electric generator to produce electricity, which is transported to a power grid over transmission lines. There are three common ways to capture geothermal steam for this process.

The dry-steam method channels geothermal steam, brought from below the earth's surface through pipes, directly into a steam turbine.

The most common method, known as flash steam, brings high-pressure, high-temperature (over 180°C) geothermal hot water to the earth's surface, where it enters a low-pressure chamber and rapidly vaporizes, or "flashes," into steam due to the sudden decrease in pressure. (Water exists as a liquid in very high temperatures underground, where the pressure is much greater than at the surface, so it vaporizes into steam very quickly once it's brought up.) The steam is then directed through a turbine, which spins to drive an electric generator to produce electrical power. Once the steam has exited the turbine, it is either released into the atmosphere as water vapour, or it is cooled and condensed back into liquid water to be injected back underground. This method is illustrated in figure 5.3.

The third approach, known as binary cycle, involves bringing lower-temperature geothermal hot water to the surface and running it through a heat exchanger that transfers the heat from the water to a working fluid (usually refrigerant) that boils at a relatively low temperature. The fluid vaporizes into a gas that is fed into a turbine to generate electricity.

To provide space heating and hot water for human consumption, geothermal resources are typically used to first

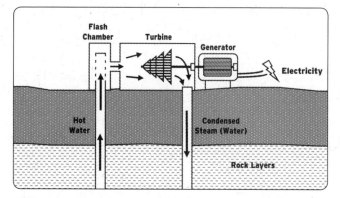

FIGURE 5.3 >>> Geothermal Flash-Stream Process

produce electricity, as described above, which is then used to provide heat and hot water via electric heaters. However, some geothermal hot water, which is not hot enough to produce steam to generate electricity, may directly provide space heating to nearby buildings by using a heat exchanger to transfer heat from the water to air and distributing it through the building, or the hot water can be used in radiant heaters. Due to the cost of insulated pipe, it is usually not feasible to transport hot water to buildings that are a long distance from a geothermal source.

Real-Life Applications

Small-scale geothermal plants are usually described as those with less than 5 MW of capacity. Since the upfront capital costs of geothermal plants require a certain critical mass or economy of scale to make them viable, these small plants are uncommon and usually need government assistance.

An example of a successful small-scale geothermal operation can be found in Altheim, Austria,[5] a municipality in the upper Austrian region with approximately 5,000 residents in an area of about 22 square kilometres. In 1989, the municipal government decided to construct, for environmental reasons, a geothermal district heating system to be fed by 106°C geothermal hot water flowing from an aquifer about 2,300 metres below the surface. Since then, the operation has been expanded to include a geothermal binary-cycle system providing up to 1.0 MW of electrical power, so the system is now providing both electricity and heat to the municipality.

Due to the capital costs of building a geothermal plant, most operations comprise dozens or even hundreds of wells. In the western United States, particularly California and

This geothermal power station at Krafla, Iceland, is one of many in the volcanically active country. Michael Utech /iStockphoto.com

Nevada, geothermal energy is used in large-scale power operations. For example, a system of 15 geothermal power plants in northern California, known as The Geysers, draws steam from wells located in the Mayacamas Mountains and has a generating capacity of about 725 MW of electricity—enough to power 725,000 homes.[6] Providing electricity to Sonoma Lake, Mendocino, Marin and Napa counties, The Geysers is one of the largest geothermal power operations in the world and the largest producer of geothermal power in the United States.

Benefits

Geothermal energy systems do not create any pollution, although they may occasionally release gases from deep inside the earth that can be easily contained. Furthermore, a geothermal plant has a smaller footprint (requires less land space) than other types of power plants, even though it generates the same amount of power.

Since there is no need to buy, transport or clean up any fuel, the costs of operating geothermal plants are relatively low. The *Renewables 2011 Global Status Report* estimates the typical energy costs related to large-scale geothermal plants to be approximately 4–7 US cents/kWh, which is relatively low among renewable energy types. Geothermal sources provide energy all or most of the time, so they offer stable, predictable energy, which translates into stable pricing.

Challenges

Geothermal sources are not typically located close to communities or transmission lines; therefore, there can be a big price tag and environmental impact involved in building access roads to the geothermal power-station site and constructing and connecting transmission lines from there to the power grid.

Using current drilling technology, we are only able to harness geothermal resources located relatively close to the earth's surface. Furthermore, geothermal-project drilling costs are expensive and increase exponentially with depth.

Although the operating costs of geothermal plants are relatively low, it takes a significant upfront capital investment to build a plant and drill boreholes, so geothermal projects usually require some government funding or policy intervention to make them financially feasible.

New Developments

Hot dry rock (HDR) technology offers access to one of the world's great untapped energy resources lying right beneath our feet. It has been estimated that there is enough heat in the earth's layers of hot dry rock that are reachable with today's drilling capabilities to supply all the energy needs of the world. HDR technology injects water down boreholes into the hot dry rock layers several kilometres under the earth's surface. Contact between the water and hot rocks produces high-pressure steam, which is extracted back to the surface to be used in a steam turbine to generate electricity.

An HDR geothermal power plant can be located nearly anywhere it is possible to access hot rock within the earth by drilling. This allows for power production in areas where it is inconvenient for other geothermal technologies to be implemented.

One drawback of HDR geothermal installations is that they may cause earthquakes if located too close to tectonic activity. In 2006, at a new HDR geothermal plant in Basel, Switzerland, an earthquake rated 3.4 on the Richter scale occurred only eight days after the plant began injecting water. Although this area had frequent earthquakes, it was determined that the plant had played a role in setting off this particular quake since its epicentre was found to be exactly at the bottom of the injection borehole. This risk can be easily mitigated by locating plants away from areas with the highest likelihood of seismic activity. HDR systems are currently being developed and tested in France, Japan, Germany, the United States, Switzerland and Australia.

KEY CHARACTERISTICS OF GEOTHERMAL ENERGY

Renewable? Yes, geothermal sources are renewable since they are naturally occurring and can continue indefinitely.

Sustainable? Yes, geothermal sources are sustainable since the geothermal energy captured for use by the current generation does not negatively affect future generations.

Provides Electricity? Yes, geothermal sources are used to generate electricity by directing naturally occurring high-pressure geothermal steam into a steam generator to produce electricity.

Provides Heat? Yes. It's possible for naturally occurring geothermal hot water to be conditioned and then used directly for either domestic hot-water or space heating, but this is only feasible if the geothermal source is close to the buildings needing the heat. Typically, geothermal energy is used to create electricity, which is then used to provide heat and hot water via electric heating systems.

Delivers Constant Energy? Yes, geothermal power plants can deliver continuous, non-fluctuating energy.

Widely Available Supply? No. Geothermal sources are currently only accessible for human use in certain regions.

Easily Transported and Stored Supply? No. Geothermal resources are not easily transported and stored; therefore, power plants are constructed at geothermal sites.

GEOEXCHANGE (GROUND-SOURCE) ENERGY

Geoexchange systems provide central heating and/or cooling to buildings using heat extracted from the ground or a body of water that are warmed by the sun. Geoexchange systems can be installed anywhere there is enough space and appropriate conditions to support an underground looped piping system. Homeowners typically install the piping systems in their backyards, while large-scale district energy projects often use community lands, such as playing fields. Lakes and oceans can also be used as a heating source, reducing the cost of excavation and drilling, but the cost of installing piping makes this practical only for homes and businesses located near these bodies of water.

Although it may seem there is not enough warmth in the ground or in a body of water to heat a building, there is actually more than enough heat to feed into a geoexchange system, which will then upgrade that heat to the required temperature.

How It Works

Geoexchange systems use plastic pipes installed underground or underwater in a loop, either horizontally or vertically, connected to a heat pump inside a building. A mixture of water and liquid antifreeze is circulated through the pipes, allowing the fluid to be warmed by the ground as it passes through the pipes and back to the heat pump.

Inside the heat pump, a heat exchanger transfers the thermal energy from the liquid antifreeze to a liquid refrigerant, which has a relatively low boiling point. The refrigerant boils to create a gaseous vapour, which is then compressed in order to increase its temperature to the desired level suitable for space heating—the more the vapour is compressed, the higher its temperature rises. Another heat exchanger inside the heat pump transfers the resulting heat either to air that is distributed throughout the building via ducts, or to water in a hot-water storage tank, which feeds it into a radiant floor system for space heating while supplying domestic hot water. In some cases, there are two heat exchangers—one transferring heat to air for space heating and the other transferring heat to water for domestic hot water.

During hot weather, the geoexchange system process is reversed to provide cooling. The heat is extracted from the building and carried back through the system to be either discarded in the ground or fed into another system to provide heating for something else such as a swimming pool.

The resulting air is often referred to as "conditioned air," since it has been heated or cooled to a comfortable interior temperature. Figure 5.4 illustrates how a typical geoexchange system works.

Real-Life Applications

Many homeowners and small businesses are opting for geoexchange systems to provide heating and cooling for their houses and buildings. Residential geoexchange systems use established

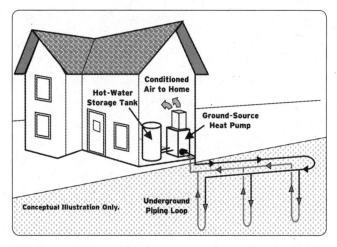

FIGURE 5.4 >>> A Typical Geoexchange System

and well-understood technology and are fairly straightforward to install in locations with the right geology. The most expensive and difficult part of constructing the system is drilling holes in the ground for the underground pipes, especially in rocky terrain. Pipes are usually installed horizontally if there is sufficient land area; however, they may be installed vertically in cases where there is limited space and the ground is fairly easy to drill into.

Geoexchange systems are also being implemented to provide cost-effective district heating and/or cooling for large-scale residential communities and business complexes. A utility company or developer could install geoexchange for an entire suburban neighbourhood, a large apartment building complex or a group of institutional buildings such as schools and hospitals. Commercial operations, including factories, retail stores and office buildings, also use geoexchange to reduce energy costs and decrease their impact on the environment. There are more than half a million commercial-scale geoexchange installations in North America today.[7]

A large-scale geoexchange system can be found at the South-wind real-estate community in Kelowna, British Columbia. The development consists of two low-rise, multi-family buildings, a separate residents' club with a pool and hot tub, and 15 townhomes. The geoexchange system at Southwind provides the community's space heating and cooling, domestic hot water, and heating for the pool and hot tub. Using a geo-exchange system has significantly reduced both the amount of energy consumed and the greenhouse gases produced by this development. Community-heating systems such as this are becoming more common due to their cost savings and environmental benefits.

Another large-scale application of geoexchange technology can be found in Toronto, where a lake loop system provides cooling to the downtown business district during hot summer days. The system cools 3.2 million square metres of office space from Queen's Park to the waterfront, using the deep water of Lake Ontario as its source of cold water. It enables the city to reduce its electricity demand by approximately 3 million kWh annually and reduce its carbon-dioxide emissions by about 732 tonnes per year.[8]

Benefits

Geoexchange systems use a small amount of heat from the ground or a body of water to heat a whole house to a comfortable temperature. The technology applies the laws of thermo-dynamics to generate heat by compressing a gas (in this case the gaseous vapour from the boiling refrigerant). Since relatively little energy is required to compress gas in order to raise the temperature, the energy efficiency of this process can reach 400 percent or higher; that is, for every unit of electrical energy that goes into operating the heat pump (to compress the gas), four units of thermal energy are delivered for space heating. This is a very good level of energy efficiency compared to a typical gas furnace, which has an energy efficiency of about 80 percent.

Challenges

The most common complaint about geoexchange is the rela-tively high expense of drilling and excavation to install the underground piping system. The payback for an installed geo-exchange system depends on these expenses and other variables, including the local cost of energy (natural gas, oil, etc.) for heat-ing and air conditioning that is offset by the geothermal heating/ cooling system. In areas where local energy costs are relatively low, it will take a homeowner longer to recoup installation costs than it would in areas where energy costs are higher.

Generally speaking, the payback for a geoexchange system is more favourable if the system is installed at the time of con-struction rather than being retrofitted into existing buildings.

This is partly because there are cost savings from doing the excavation for the pipes at the same time as the excavation of the building foundation, and from installing the correct ducting at the time of construction rather than having to modify it during the retrofit.

Those who find the upfront costs of geoexchange systems too expensive may choose to use air-source heat pumps instead. They work much like geoexchange heat pumps, except they use heat from the outside air instead of the ground. Air-source heat pumps are not as energy-efficient as geoexchange systems, and in cold climates their efficiency further decreases due to there being even less thermal energy (heat) in the air. Therefore, they consume more energy than geoexchange systems to maintain the same temperatures.

New Developments

The 2010 Winter Olympic Village in Vancouver implemented a quasi-geoexchange system that uses the village's sewer system, instead of the ground, as a heat source. This new technology taps renewable, locally available municipal liquid waste (sewage) as its energy source. It works in the same way as a ground-source system, except it feeds liquid waste, instead of liquid antifreeze mixture, into a heat-pump system in order to generate sufficient heat for residential space heating and domestic hot water. This method is more efficient and cheaper than typical geo-exchange installations; since sewage is warmer than the ground, less energy is needed to compress the refrigerant to the required temperature. Furthermore, there is no need for expensive dril-ling and excavation. During the coldest days of the year, this system is augmented by high-efficiency natural-gas boilers.

KEY CHARACTERISTICS OF GEOEXCHANGE ENERGY

Renewable? Yes, ground heat is renewable since it is naturally replenished by solar radiation on an ongoing basis.

Sustainable? Yes, geoexchange systems are sustainable since they can be used by the current generation without negatively affecting future generations.

Provides Electricity? No, geoexchange systems do not produce electricity.

Provides Heat? Yes, geoexchange systems use heat stored in the ground to produce domestic hot water and/or space heating, and can also be used to cool buildings.

Delivers Constant Energy? Yes, energy captured from geoexchange systems can be constant, even though the temperature in the ground may vary over time.

Widely Available Supply? Yes, ground heat is widely available as long as there is sufficient land space.

Easily Transported and Stored Supply? No, the heat stored in the ground is extracted and transported only short distances. It is not easy to transport this heat across medium or long distances without significant energy loss. Warm water can be transported longer distances using insulated pipes, but they can be expensive.

chapter six

HYDRO ENERGY

Electricity can be

generated by capturing

the **ENERGY** in

FLOWING WATER.

THE BASICS

Hydro energy, currently the most widely used form of renewable electricity, refers to energy generated by capturing the kinetic and potential energy of flowing water. Ultimately, hydro energy originates from the sun, which causes water in oceans and lakes to evaporate and form clouds. Subsequently, the vapour in the clouds condenses back into water and falls to the ground as snow or rain, which often flows into rivers that can be used to generate hydroelectricity (hydropower).

By the end of 2010, the worldwide installed capacity of hydropower reached 1,010 GW, representing approximately 3.4 percent of all global energy consumption, 16 percent of the world's electricity and about 80 percent of electricity generated from renewable sources.[1] Hydropower is currently used in about 150 countries, with China, Canada, Brazil, the United States and Russia having the greatest installed capacities. China has seen the greatest growth in recent years, nearly doubling its capacity since 2005 to reach 213 GW. Canada added about 500 MW of capacity in 2010 to reach a total of 75.6 GW. The United States experienced very slow growth in the last few years due to the recession, ending 2010 with a capacity of 78 GW. Russia has about 55 GW of hydropower capacity.[2]

Hydro energy is one of the oldest types of energy harnessed by humankind. Over 2,000 years ago, the Greeks and Romans placed waterwheels along streams so the water could provide mechanical energy to turn the wheels to grind corn and wheat. The use of waterwheels spread throughout Asia and Europe and evolved to provide mechanical power to mills and other industrial machines. In the centuries that followed, engineers continually improved on the design of waterwheels to expand their capabilities. By the 18th century, the waterwheel was starting to be replaced by water turbines, which guided the water flow so that it acted with the greater efficiency on the wheel blades. Water turbines also worked better in faster-moving and less voluminous mountain creeks and waterfalls, thus expanding the use of hydro energy.

HYDROPOWER SYSTEMS

The electric generator was invented late in the 18th century for use with water turbines to produce electrical energy, and the first hydropower stations were built by the late 19th century. The world's first commercial hydropower plant began operating in 1882 in Wisconsin, with an output of about 12.5 kW. By 1886, there were about 45 hydropower plants across the United States and Canada. Nikola Tesla, famous for his contributions to the development of commercial electricity, was the first person to use hydropower plants to distribute electricity on a wide scale. In 1896, he enabled the Niagara Falls power plant in New York to distribute electricity to Buffalo. By the early 1900s, there were a few hundred hydropower plants in the United States and a growing number elsewhere. During the 20th century, many more large-scale hydropower dams were constructed around the world.

How It Works

To produce hydropower, water is channelled through turbines to rotate their blades that turn a shaft that drives an electric generator. A traditional hydropower dam holds water back in a reservoir and releases it through turbines as needed to meet electricity demands. Smaller hydro systems often redirect some of the water from a fast-flowing river through pipes to hydropower stations, so water is not stored in reservoirs.

The two main factors that determine the electricity output from hydro energy are the head (vertical fall) and flow (volume flow rate) of the water. The vertical fall is the vertical distance from the upstream level to the downstream level or, in the case of a dam, the distance from the level of the reservoir to the lower body of water at the output of the turbine. The volume flow rate is the amount of water that passes through the turbine per unit of time. The greater the vertical fall and flow, the more electricity will be generated.

Another factor that affects electricity output is the energy efficiency of the components, such as the turbines and generators. The more efficient the turbines are at transferring the kinetic and potential energy in the water to mechanical energy in the rotating shaft, the greater the electricity output achieved. Whenever energy is transferred, there is some energy loss (or inefficiency). The most modern hydro turbines operate at mechanical efficiencies greater than 90 percent.

TYPES OF HYDRO OPERATIONS

While definitions vary, hydro projects are commonly called *small hydro* if they have an installed capacity of less than 10 MW, which will supply power to approximately 5,000 homes. Typically these are small hydropower operations that use either small dams or diversion systems.

Hydro operations that divert river water, instead of using dams, are referred to as *run-of-river* because they capture energy in real time from running rivers as they flow, instead of using dams to trap water in reservoirs. Run-of-river operations divert some water from the river into an enclosed pipe called a penstock. The penstock feeds the water downhill to a power station that contains turbines and a generator. The natural flow, due to the elevation drop of the river, creates the force required for the water to spin the turbines, which generate electricity that is fed into the power grid. The water leaves the generating station and is returned to the river with minimal alteration to the existing flow or water levels. (It should be noted that run-of-river projects typically range from small- to medium-scale). Figure 6.1 illustrates a typical run-of-river configuration.

As an example, Brandywine Creek, near Whistler, British Columbia, is home to a fully functional 7.6-MW run-of-river hydro operation that produces approximately 38,000–42,000 MWh of electricity annually.[3]

Small hydro is a relatively mature type of green energy, as many such sites around the world have been active for decades. Currently about 6 percent of all hydropower comes from small hydro projects.[4]

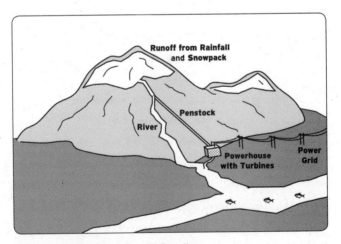

FIGURE 6.1 »» A Typical Run-of-River Configuration

Microhydro systems, usually defined as hydropower operations with a capacity of 100 kW or less, are ideally suited to generate electricity for a home, farm or small community. For example, a microhydro system could be a homemade hydropower structure installed on a stream running through a homeowner's property.

Large-scale hydropower has been produced since the late 1800s, and it now provides about 20 percent of the world's electricity, including about 60 percent of Canada's electricity and roughly 10 percent of the United States' electricity.

A typical large-scale dam, as shown in Figure 6.2, is built on a wide, fast-moving river with a significant drop in elevation. The dam holds water behind it in a large reservoir. Near the bottom of the dam wall is the water intake. Gravity causes water to

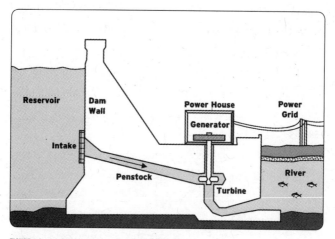

FIGURE 6.2 ››› A Typical Large-Scale Hydro Dam

RANKING	NAME	COUNTRY	2010 CAPACITY (GW)
1	Three Gorges Dam	China	21.0
2	Itaipu	Brazil, Paraguay	12.6
3	Guri	Venezuela	10.3
4	Tucurui	Brazil	8.4
5	Grand Coulee	United States	6.8

TABLE 6.1 »» The Five Largest Hydropower Dams in the World

flow through the intake and into a penstock inside the dam. At the end of the penstock is a turbine with blades that are turned by the force of the moving water. The rotating blades turn a shaft connected to a generator, which produces the electricity. Power lines are connected to the generator to carry electricity to its destination via the power grid.

Table 6.1 lists the five largest hydropower dams in the world, according to their total capacity when running at peak.

Benefits

Hydro energy has relatively low and stable operating costs, since it uses free-flowing water and is not dependent upon costly or limited resources such as uranium or oil. Once installed, hydro systems operate very reliably with minimal need for operators, compared to fossil-fuel power plants. An additional economic benefit is that hydropower systems have very long operating lives—over 100 years in some cases.

Hydro energy's cost per kilowatt is usually lower than that of wind and solar electric power, mostly because it uses well-established technology and traditional components that are less expensive than solar PV and wind-turbine technology.

Once a hydropower project is constructed, it does not produce direct waste or greenhouse-gas emissions.

Hydropower systems can be installed in a wide range of sizes to provide electricity for a single home, city or even multiple cities.

Another benefit of hydro-energy systems is that they can start up and shut down production of electricity quickly and easily compared to fossil-fuel plants. In most cases, the operators

The Grand Coulee Dam on the Columbia River in the state of Washington is the largest hydropower facility in the United States. John Beckman Jr./iStockphoto.com

merely close gates to stop water from flowing through the turbines. Furthermore, they can vary the amount of electricity being generated by using valves or gates to control the flow of water.

The amount of power generated from hydro is predictable and available on a continuous basis, although most rivers have much higher flows in the spring than at other times of the year.

Challenges

The main disadvantages of large-scale hydro energy are the upfront costs involved in constructing dams and the potentially negative environmental impacts. Although hydro-dam technology is well established and cost-effective, the sheer size of large-scale hydro plants requires a large investment.

When a dam is built, large areas upstream are flooded, potentially dislocating people, eliminating farmland and destroying bird and animal habitats. In 2011, China's central government for the first time acknowledged the negative impacts of the Three Gorges Dam, the world's largest hydropower dam. The State Council admitted that the dam had caused severe problems to the environment, shipping, agricultural irrigation and water supplies in the lower reaches of the Yangtze River, an area of 633,000 square kilometres.

Some individuals and river-sports enthusiasts are opposed to both large and small hydro projects because they divert rivers, affecting both natural and human activities by altering water flow, water temperatures and habitat. Therefore, it is important for hydro-project developers to select sites that will have minimal impact on fish, wildlife and communities.

Others argue that hydro dams flood valleys full of plants and trees that are needed to absorb carbon dioxide from the atmosphere as part of the natural carbon cycle. This is a valid concern that should be weighed against the amount of carbon dioxide emissions that are being eliminated by using hydropower instead of fossil fuel–based power.

Despite these drawbacks, hydro energy's low operational costs compared to other renewable-energy types ensure it will continue to play a key role in meeting our energy demands.

New Developments

While there are still massive hydropower projects being planned for the future, especially in developing countries, there is a growing interest among developed countries in medium-scale, run-of-river projects as an alternative to large-scale dams. However, they cannot be installed anywhere on a river; they must be situated where there is sufficient vertical drop, flow and relatively close access to transmission lines. Such projects are larger than the small-scale ones described earlier in this chapter, but still have a lower impact on the environment than traditional dams. As an example, the Toba Montrose run-of-river hydropower project in northern British Columbia became fully operational in August of 2010, after nearly three years of construction. It is expected to deliver 715,000 MWh per year of power to the British Columbia power grid over the next 35 years.[5]

There are also an increasing number of pumped-hydro storage systems being used in conjunction with traditional hydro dams. These systems help hydropower plants meet high peak demands by moving water between two water reservoirs at different elevations. At times of low electrical demand, excess electricity capacity generated by the hydropower plant is used to pump water into the higher reservoir (thereby converting electrical energy back into potential energy). When demand increases, the water in the higher reservoir is released (converting potential energy into kinetic energy) and channelled through a turbine to produce electricity again as it flows to the lower reservoir. This not only helps the plants deliver electricity to consumers when they need it, but it also enables them to be more profitable by using more of their power-producing capacity. By the end of 2010, the world's total pumped storage capacity was 136 GW, with the vast majority being in Europe, Japan and the United States. Another 5 GW of capacity was under contract by early 2011, and the market is expected to rise about 60 percent over the next five years.[6]

There are ongoing improvements being made to hydro turbines. For example, computer programs are being developed to study the dynamic characteristics of hydraulic systems and mechanical designs of hydro turbines.

Advancements are being made in the design of reversible-pump turbines, which are used in pumped-storage systems. Reversible-pump turbines can operate in one direction as pumps and in the other direction as turbines. They are a significant development in the field of modern hydropower engineering, as they offer much greater efficiency than the current pumped-storage turbines, which generate electricity to run an electric motor to drive the pumps.

In a new twist on turbine development, a Norwegian company has developed technology that uses the excess pressure from water running through municipal water-distribution systems to drive turbines and create electricity. They are essentially replacing pressure-reducing valves with very small turbines to generate 80–300 kilowatts of power, which can then be fed into the power grid or consumed locally.[7]

KEY CHARACTERISTICS OF HYDRO ENERGY

Renewable? Yes, all hydropower is considered renewable energy because it is generated from the flow of water, which is continuously replenished in nature as long as there are rivers.

Sustainable? Run-of-river systems are considered sustainable since the energy we capture from them for the current generation has minimal impact on future generations; however, large hydro dams are arguably less sustainable, since they damage the local environment and affect future generations by taking away their opportunity to enjoy the river and natural habitat as it was before dam construction.

Provides Electricity? Yes, flowing water is used to generate electricity by directing the water through turbines, allowing the water to rotate the blades, which turn the shaft of an electric generator.

Provides Heat? Not directly, although the electricity generated by a hydropower system could be used to power an electric heater for space heating or hot water.

Delivers Constant Energy? Yes, although the flow of water in rivers and streams is not absolutely constant, it is steady enough to provide for constant hydro energy when it is needed. The supply of hydropower is easily controlled and typically delivered according to the demand level of consumers.

Widely Available Supply? Yes, relatively speaking. Rivers and streams are found in every country, although their supply is more abundant in some regions than others, and some rivers have better head and flow than others.

Easily Transported and Stored Supply? Yes, water can be easily transported or pumped uphill through pipes to a desirable location where it can be directed into a power station or stored in a reservoir.

chapter seven

OCEAN ENERGY

Ocean energy can be

harnessed from the power

found in ocean **TIDES**

and **WAVES**.

THE BASICS

Covering more than 70 percent of the earth's surface, oceans absorb and retain massive amounts of the sun's energy. If humankind could harness just a fraction of that energy, it would be enough to satisfy the power needs of the world. The total potential of ocean energy has been estimated by the International Energy Agency's Implementing Agreement on Ocean Energy (IEA-OES) as between 20,000 and 90,000 TWh/year, which is more than the total amount of electricity consumed in the world.[1]

Two types of ocean energy—tidal and wave—can be harnessed to provide electrical energy through various methods such as those shown in figure 7.1. Although tidal and wave energies both originate in ocean waters, they are generated from different sources: tides are caused by gravitational forces, while ocean waves result from wind passing over open bodies of water. Each type of ocean energy employs various power-generating technologies.

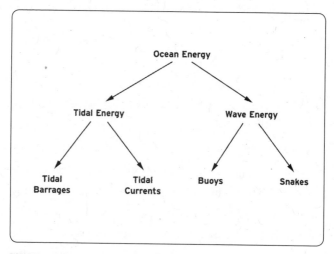

FIGURE 7.1 ››› Typical Types of Ocean Energy

TIDAL ENERGY

Tides represent the cyclic rising and falling of ocean levels caused by the gravitational pull of the moon and the sun. The total global potential for tidal energy is immense, with estimates exceeding 450 TW.[2] There are two fundamentally different approaches to capturing tidal energy: *tidal barrage* and *tidal current*. Tidal-barrage energy resources are generally best in areas where the water depth is relatively shallow and there is a wide range in tide levels. Tidal-current energy can be captured as tidal water rushes through narrow channels; therefore, the best tidal-current energy resources are found in regions with such passages.

How It Works

As mentioned above, there are two main ways to capture tidal energy. The first and more common approach, tidal barrage, involves building a barrage, similar to a dam, across the mouth of a bay or an estuary with relatively shallow water and a large tidal range. The barrage has gates that open and close to control the water flowing through it. As the tide is rising the gates are held open, allowing the basin to fill with water. When the tide reaches its highest point, the gates are closed to trap the water in the basin, creating a water reservoir. As the tide goes out, there is a growing difference in elevation between the reservoir level and the sea level, and this produces sufficient head to use for hydroelectric production. Once sufficient head exists (while the tide is still falling) the gates are opened to allow water to flow from the reservoir to the sea through the barrage tunnels, which contain turbines connected to a generator to produce electricity. The gates remain open when the tide is at its lowest level and while it begins to rise again, allowing the basin to refill and repeat the cycle. Figure 7.2 illustrates the configuration of a tidal-barrage power station.

The second approach, still mostly in development, uses turbines to harness the kinetic energy of tidal currents rushing through narrow passages, and is referred to as tidal-current energy or *tidal-stream energy*. This approach focuses on tidal currents found in regions with high tidal ranges and narrow passages such as straits, narrows or fjords. Tidal-current energy systems typically use submerged turbines located in narrow, shallow channels; as tidal currents rush through the passages,

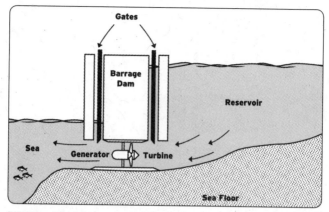

FIGURE 7.2 >>> A Tidal-Barrage Power Station

the water spins the turbine blades, which drive a generator to produce electricity. Multiple tidal turbines are sometimes referred to as a *tidal-stream farm*, similar to a wind farm.

Underwater turbines work on the same principle as wind turbines by using the kinetic energy of moving water (instead of moving air) and converting it into electrical energy. The velocities of the currents are usually lower than those of the wind; however, because water is denser than air, the turbines can be smaller than their wind counterparts for the same installed capacity. Turbines can be rigidly mounted in the seabed, pile-mounted, semi-submersible with moorings, or attached to a floating structure. A conceptual illustration of a tidal-current energy system is shown in figure 7.3.

Real-Life Applications

Tidal-energy systems are not typically used for residential purposes; rather, they are often medium-to-large developments

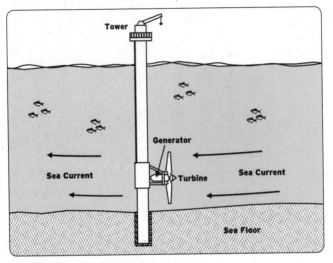

FIGURE 7.3 >>> A Conceptual Tidal-Current Energy System

managed by private companies or utility companies. The world's first small-scale commercial tidal-current turbine to generate electricity for the grid was commissioned in Northern Ireland's Strangford Lough. The 1.2-MW SeaGen plant produces enough power for about 1,500 homes.[3] Other small-scale (less than 10 MW) commercial installations represent the first phases of planned larger projects. In 2009, South Korea completed the 1-MW Jindo Uldolmok tidal-current energy project, which is the first phase of a planned 90-MW plant to be completed by 2013.

Today a handful of commercial tidal-barrage systems are generating power, and another handful are under construction. Unlike tidal-current energy systems, tidal barrages have been in use for decades. The Rance Tidal Power Station, operated by the largest utility company in France, is the world's first and largest tidal-barrage power station. It commenced operations in 1966 and is still operating today. Located on the estuary of the Rance River, the facility uses a barrage constructed across the estuary. As the tide goes in and out, ocean water flows through tunnels in the barrage and turns turbines to generate electricity. This site was chosen because of the wide range between low- and high-tide levels, averaging 8 metres and reaching up to 13.5 metres. The facility's installed capacity (the amount of electricity generated during peak times) is about 250 MW, and the total annual output is approximately 550 GWh.[4]

The only tidal-barrage power station in North America is the Annapolis Royal Generating Station in Nova Scotia, which generates just 20 MW. Opened in 1984, the station exploits the tidal difference created by the large tides in the Annapolis Basin using a tidal-barrage system similar to that described earlier in this chapter.

Benefits

Tidal energy uses an abundant source—ocean water—that is available day and night. Since tidal schedules are known and predictable, tidal energy can produce reliable power.

Installed tidal-energy systems are non-polluting and produce no greenhouse gases. Unlike land-based wind turbines, tidal-energy turbines are quiet and out of sight, since they are mostly submerged and located away from homes and businesses.

Challenges

The main challenges for tidal-barrage power stations are that they are expensive to build and have detrimental impacts on habitat both upstream and downstream. Tidal-barrage facilities change water levels and sedimentation and alter the amount of time that estuary mud flats are exposed, thereby affecting fisheries, birds and other wildlife. They can also trap marine life when the gates are closed. The Annapolis Royal Generating Station in Nova Scotia has trapped at least two whales in its basin since it commenced operations, and one of those whales died as a result.

The key challenge for tidal-current energy systems is that they are difficult to install and maintain because they are usually installed in treacherous, narrow channels with fast currents rushing through them.

Both tidal-barrage and tidal-current energy systems are subjected to tides that are predictable, but they are not continuous, so they deliver only intermittent electricity. Also, only certain locations have sufficient tidal ranges for producing tidal energy, and the tides cannot be transported, so the power-generation infrastructure must be located near them.

New Developments

Tidal-current turbines are still an immature technology, and a number of designs are being researched, prototyped and tested. Some designs feature rotor blades that can be pitched through 180 degrees to allow them to operate in bidirectional flows during ebb and flood tides.

Due to the difficulty with installing turbines in treacherous rushing waters, there are ever-evolving ideas for new ways to install and maintain submerged turbines. For instance, some designs feature a vertical steel tower with a sliding turbine system, which can be brought to the surface whenever it requires maintenance, providing safe, easy access for maintenance staff.

WAVE ENERGY

Waves are formed by wind passing over the surface of the ocean. A wave's energy is proportional to the square of its height; therefore, a three-metre-high wave has nine times more energy than a one-metre-high wave. The best ocean-wave energy resources are generally found where the strongest winds are found—within 40 to 60 degrees of latitude.[5] The largest waves are formed in regions where winds blow long distances over water in the same direction, often on the west coasts of continents. Areas particularly rich in ocean-wave energy include the western coasts of Scotland, northern Canada, southern Africa, Australia and the northeastern and northwestern coasts of the United States.[6]

How It Works

The two most common types of technologies currently being used to extract wave energy are known as snakes and buoys.

There are various types of ocean-wave snake technologies under development. One type is comprised of a series of long cylindrical tubes, connected by hinged joints, floating on the surface of the water and anchored on one end, as illustrated in figure 7.4. As waves move the joints up and down and side to side, hydraulic rams drive an electric generator located between the joints. The electric generator may be connected to batteries that store the energy generated from the system, or to a subsea power cable that transports the electricity to shore, where it is connected to a power grid.

Ocean-wave buoys are still in the prototype and testing stages and have not been put into commercial use yet. Like snakes, they use floating devices to capture energy from rising and falling waves.

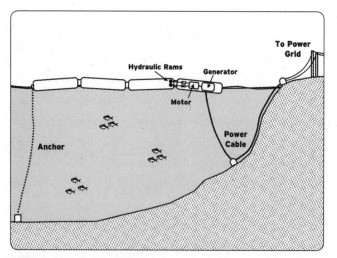

FIGURE 7.4 >>> An Ocean-Wave Snake

Real-Life Applications

Like tidal energy, wave energy is not being planned for small-scale applications such as residential or small-business systems. The small-scale projects that have been installed typically represent the initial phase of a larger project. In May 2010, Scotland launched a 600-foot-long ocean-wave snake. This 750-kW device is the first of 66 planned machines as part of a *wave farm*. The device will undergo several years of testing before commercial use.[7]

There are no installed large-scale ocean-wave projects today, reflecting the fact that this is still an immature industry. The Aguçadoura Wave Farm in Portugal was the world's first attempt at a commercial wave farm. Located five kilometres off the coast, the farm used three ocean-wave snake devices. The farm was officially opened in September 2008 and initially operated with a capacity of 2.25 MW. It shut down a few months later due to technical problems with some of the bearings in the devices; however, a solution has since been found. The second phase of the project was planned to increase the installed capacity to 21 MW, using 25 more devices, but due to the financial crises that hit Portugal and many other countries in 2008, sufficient funding has not been found for the next phase. Nevertheless, the project has proven that wave energy can be successfully harnessed and transmitted to shore and into the power grid.[8]

Benefits

Installed wave-energy systems are non-polluting and use a free source of ocean water that is continuously replenished in nature.

In some parts of the world, waves are extremely powerful and virtually constant.

In addition to emitting no greenhouse gases, wave-energy systems typically have a low impact on surrounding marine life if designed properly. Ocean-wave technologies also tend to be relatively undetectable to humans, since they are located off-shore and lie under or along the surface of the water.

Challenges

Ocean-wave energy technologies are still in the early stages of development, and there are few commercial wave-energy proj-ects in operation. According to some estimates, ocean energy may be about three to six years behind wind power in terms of commercial rollout. The main challenges include harsh ocean conditions, the upfront costs of research and development, and the inability to produce electricity when the ocean is calm, which is difficult to predict.

Ocean waves are intermittent, since they depend on the wind, which is also intermittent, so they cannot deliver continuous energy. It's impossible to transport waves to other locations, so power-generation systems must be located near the ocean. Another issue is the lack of access to power grids near coastal or offshore ocean-energy projects. To address this problem in Europe, the British government has agreed to provide funding for the development of a "Wave Hub" off the northern coast of England, which will act as a giant extension cable, allowing wave energy–generating devices to be connected to the European electricity grid. The Wave Hub is like a socket sitting on the seabed for wave-energy devices to plug into. A cable from the Wave Hub to the mainland will transmit electrical power produced from the wave-energy devices to the electric grid. The Wave Hub will initially allow 20 MW of capacity to be connected, with potential expansion to 40 MW.

Wave-energy systems are exposed to risks associated with extreme weather. In 2010 a massive swell at Port Kembla, 150 metres off the coast of Australia, sank the country's first ocean-wave system, which was feeding power to the Australian electricity grid. Although the failure was with the moorings, not the wave-energy technology, this incident points out that wave-energy developers are still learning how to build systems that can withstand extreme weather conditions.

New Developments

Wave-energy technology has benefitted from both government and private funding, especially in the United Kingdom, Ireland, Portugal, Denmark, France, Australia, South Korea, Canada and the United States. Currently the United Kingdom is the largest market for ocean-energy technologies and has at least 20 companies developing prototype equipment.

Ocean-wave technology development will focus on reducing the costs of these systems in order to make them more competitive with other renewable-energy technologies. Efforts will also be made to design the systems to be more durable in harsh ocean conditions, including developing better ways to moor the ocean-wave devices.

OTHER METHODS OF GENERATING OCEAN ENERGY

Diverse new technologies are being developed and tested to exploit different ocean resources. *Salinity gradient* and *ocean thermal-energy conversion* (OTEC) are two of those new technologies that are still in the research and development stage.

Salinity-gradient power technology uses osmotic pressure between water bodies of different salinity. (Osmotic pressure refers to the pressure exerted by the flow of water through a semi-permeable membrane separating two solutions with different concentrations.) One method involves two liquids— salt water and fresh water—separated by a semi-permeable membrane. The fresh water diffuses through the membrane in order to dilute the saltwater solution. The resulting flow of water drives turbines connected to a generator, which produces electrical energy. At this time, the technology is not ready to produce power commercially; however, the first salinity-gradient prototype installation has been operational since December 2009 in Norway.

OTEC technology employs a complex process that uses temperature differences between surface sea water and deep sea water to drive steam turbines to generate electricity. Ammonia (a refrigerant that boils at a very low temperature) is circulated through a closed-loop piping system. Warm surface sea water circulating in an adjacent piping system provides enough warmth to cause the ammonia to boil and turn into a pressurized gaseous vapour that turns the blades of a turbine. The ammonia gas continues to circulate through its closed loop piping system, passing another adjacent piping system containing cold water pumped up from the ocean depths. Exposure to the cold water causes the ammonia gas to cool and liquefy; it is then pumped around the loop toward the warm sea water that will once again vaporize it in a continuous cycle. OTEC technology requires the use of large power plants because of low thermal efficiency; hence, a large capital investment is needed for such plants. Only small-scale versions of OTEC have been constructed, including one in Japan that can create 100 kW of electricity. Another small-scale system off the coast of Hawaii produces 50 kW of electricity.

KEY CHARACTERISTICS OF OCEAN ENERGY

Renewable? Yes, ocean energy is renewable because ocean water is continually replenished in nature and the energy source (the sun) is available indefinitely. As long as the moon and sun exist, ocean energy will always exist.

Sustainable? Yes, ocean energy is mostly sustainable since it can be used by society today with minimal negative impact on future generations. Ocean energy systems may affect ocean habitat, but so far these effects are reported to be minimal.

Provides Electricity? Yes, ocean energy is captured to generate electricity, most commonly by either forcing tidal currents through turbines that drive an electric generator or by capturing energy from up-and-down wave action.

Provides Heat? Not directly. The electricity produced from ocean energy may be connected to an electric heater to produce space heating or hot water.

Delivers Constant Energy? No. Although tidal-current energy can deliver predictable electricity based on known tide patterns, it does not deliver constant energy due to the ebb and flow of tides. Ocean-wave energy fluctuates with weather conditions.

Widely Available Supply? Yes, ocean tides and waves are widely available around the world, although not all countries border oceans, and not all areas of the ocean are suitable for tidal- and wave-energy systems.

Easily Transported and Stored Supply? No, ocean tides and waves cannot be transported or stored. Ocean-energy technology captures energy on or in the ocean in real time, so electricity-generating plants must be located at those sites. After ocean energy has been converted to electrical energy, the resulting electricity can be transported via power lines.

chapter eight

BIOMASS ENERGY

Biomass energy is a

critical element in the

alternative-energy mix

that can help society

REDUCE its reliance

on **FOSSIL FUELS**.

THE BASICS

Biomass energy, otherwise known as *bioenergy*, is any type of energy derived from biological matter, known as biomass, including organic material such as wood, agricultural crops and human or animal waste.

Biomass energy has been used by humans since early man started burning wood for heat to stay warm and cook food. Until the 19th century, when society realized the full potential of fossil fuels, wood was the predominant fuel.

Currently, biomass energy is the most commonly used type of renewable energy and is also one of the fastest growing due to the wide availability of biomass such as wood waste from the forest industry, animal waste from the agriculture industry and human waste found in landfills and waste water. In 2010 approximately 69 percent of the world's renewable energy consumption and roughly

11 percent of its total energy consumption was provided by biomass energy.[1]

Biomass is *carbon neutral* since it removes as much carbon dioxide from the atmosphere over its lifetime as it puts into the atmosphere. Biomass material is derived from living or recently living organisms such as trees, plants, animals, vegetables and grains. These biological organisms absorb carbon dioxide as they grow and release it when they die and decay, or burn, as part of the natural carbon cycle, thereby releasing roughly the same amount as they absorb over their lifetime. Therefore, there is a balance between the amount of carbon that is emitted and absorbed by biomass, and no "new" carbon is added to the atmosphere during this process. In contrast, when humans extract fossil fuels and use them for energy, they

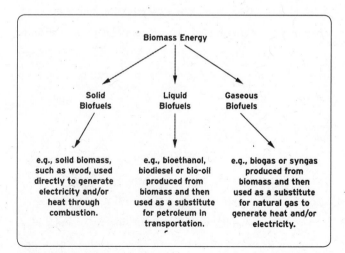

FIGURE 8.1 >>> Types of Biomass Energy

emit "new" carbon into the atmosphere that would otherwise stay locked in the ground, thereby leading to an overabundance of carbon in the atmosphere. For this reason, fossil fuels are not carbon neutral.

Biomass energy can be generated from three types of biomass fuels (biofuels): solid, liquid and gaseous. Figure 8.1 illustrates these three types and provides examples of each.

SOLID BIOFUELS

Solid biofuels include solid biomass, such as wood, that can directly generate heat and electricity through combustion. This is the oldest method for generating heat for cooking and heating and is still the most common way to generate energy from biomass today. The idea of combustion often conjures up a negative image of old-fashioned incinerators that emitted massive amounts of air pollution; however, today's combustion systems are much more sophisticated and are built with technology to maximize efficiency and minimize air emissions.

How It Works

A typical modern biomass combustion system uses a specially designed furnace that limits air emissions as it generates heat to be used for space heating or for boiling water to create high-pressure steam that drives a turbine to generate electricity. Waste wood—pellets, bark, sawdust, tree trimmings, construction pallets or insect-damaged wood—is often used as the solid biofuel for this process.

Real-Life Applications

Homeowners carry out small-scale combustion of wood and wood waste by installing wood-burning stoves to generate heat for space heating. However, to ensure the process is efficient and clean-burning, it is important to use a stove that is built for this purpose and make sure it meets the standards set by local building bylaws and home-insurance companies.

Open wood-burning fireplaces are not ideal for space heating because they are not efficient (about 90 percent of the heat goes up the chimney) and result in high levels of emissions that contribute to air-pollution problems outside the home and air-quality problems within the home.

Wood pellets are a common type of solid biofuel. Miodrag Nikolic/iStockphoto.com

For commercial applications, some industrial facilities with access to plentiful, inexpensive wood are combusting it to provide energy for their operations. Some pulp and paper mills combust their own wood waste to generate heat and electricity for their buildings and machinery.

A number of countries, especially in Europe, are retrofitting coal-fired power stations into biomass power stations by feeding wood pellets into their coal burners. Other countries, like the United States, which burns about a billion tonnes of coal per year,[2] are slower to make the move from coal to biomass.

Due to state-of-the-art combustion technologies that minimize pollution, some new residential developments combust wood waste, as an alternative to fossil fuels, to provide community energy.

Municipal waste-to-energy facilities often use combustion systems to turn solid waste into electricity. Using incineration to convert municipal solid waste to energy is not a new concept and usually involves burning garbage to boil water, which creates high-pressure steam to spin the blades of a turbine that drives a generator to produce electricity. Nowadays, these systems are much more efficient and are required to meet stricter air-emission standards than in the past. One of the benefits of combustion systems for municipalities is that they burn all types of garbage, including organic matter, plastics and synthetics. This eliminates the need to separate out organic matter; however, this also means they miss the opportunity to capture and use biogas from the organic material. Other waste-to-energy methods such as anaerobic digestion, which is discussed later in this chapter, produce biogas and can be more environmentally friendly.

Benefits

From a financial point of view, solid-biomass energy is one of the most feasible and profitable types of alternative-energy options and is already in commercial use in many countries. Bioenergy can help organizations achieve savings by using inexpensive waste biomass for fuel instead of fossil fuels. The use of solid biomass can also help stimulate local employment and leverage local biomass waste. Solid biomass is also carbon neutral, and can be replenished indefinitely by replanting trees and plants.

Challenges

A key challenge for solid-biomass energy is that it typically requires a significant quantity of biomass that has to be transported to a central processing facility; therefore, there are sometimes logistical issues in stockpiling sufficient quantities in one location for energy production. With the increasing popularity of biomass energy, it is possible there may even be a future shortage of biomass to fulfill demands.

There is also the perception that cutting down trees for fuel is bad for the environment. In the past, forests were cut down without being replanted or properly managed. However, as long as trees are replanted and farmed responsibly, they can provide an endless source of biomass supply for renewable energy.

Another common concern regarding solid-biomass combustion is the air pollution it creates. While some developers are successfully building state-of-the-art combustion facilities to provide energy in new communities, others are facing strong objections from nearby residents who are very concerned about air quality. However, if the latest technology is used in combustion facilities, air emissions can be minimized to meet or exceed government requirements.

New Developments

Scientists are investigating numerous ways to replace coal, the dirtiest of fossil fuels, with solid biofuels having similar energy properties to coal. There have been some studies carried out to explore the possibility of replacing coal with bio-char, which is charcoal produced through *pyrolysis* (chemical decomposition of a material due to heat in the absence of oxygen) of solid biomass; however, this is still in the development stage. Further along in development, waste-water treatment biosolids are already being used as a biofuel to replace coal in some applications.

LIQUID BIOFUELS

Produced from biomass, liquid biofuels are used as transportation fuels to replace petroleum. Global transportation, including planes, trains, ships, trucks and automobiles, represents an area of growing fossil-fuel consumption that drives up fuel prices and causes greenhouse-gas emissions. Finding new sources of renewable fuels for transportation is vital in reducing the financial and environmental costs of fossil fuels. Liquid biofuels, especially bioethanol and biodiesel, could play an important role.

Ethanol, also called *ethyl alcohol*, is a volatile, flammable, colourless liquid best known as the type of alcohol found in alcoholic beverages. It can be produced chemically from ethylene or biologically from the fermentation of certain agricultural crops. When it is produced from biomass, it is often referred to as *bioethanol*.

Biodiesel is a clean-burning, non-toxic and biodegradable renewable fuel derived from organic matter that is used as a replacement for or additive to petroleum diesel.

How It Works

To produce bioethanol, arguably the dominant biofuel in the world today, starch crops such as sugar cane and corn are transformed into sugars that are fermented, distilled and then dehydrated into bioethanol. The resulting liquid bioethanol is typically used as fuel for vehicles. Bioethanol is used in various concentrations ranging from E10 (meaning 10 percent ethanol and 90 percent gasoline) to E100 (meaning 100 percent ethanol).

Biodiesel is produced by subjecting either animal fat or vegetable oil to a chemical process called *transesterification*, followed by separation and purification. It is often mixed with traditional petroleum diesel (which is made from refining oil).

For example, B40 biodiesel means it contains 40 percent bio-diesel and 60 percent petroleum diesel. Rapeseed is the most common ingredient of biodiesel production in Europe, while soy oil is most common in the United States. Palm oil, canola oil, sunflower oil and jatropha oil (made from the seeds of *Jatropha curcas*, a flowering plant found mostly in Mexico and Central America) are also used. Animal fats and leftover cooking oils can be used but first require costly conditioning to remove contaminants. Biodiesel fuel has the same viscosity (thickness) and density as petroleum-based diesel, so it can be easily sub-stituted into any diesel engine. Biodiesel has a higher freezing point than petroleum-based diesel so is more prone to solidify in cold temperatures. As a result, the ratio of biodiesel to petroleum diesel in a blended diesel fuel will be lower in cold climates.

Real-Life Applications

It's possible to make bioethanol at home as a fuel for personal vehicles and farm equipment, although it's not as easy to make as homemade biodiesel, and it's difficult for inexperienced pro-ducers to achieve a consistent grade. Recipes for homemade bioethanol are available on the Internet.

Anybody can make biodiesel in their home from new or used vegetable oil, both of which are relatively cheap. Using second-hand oil is not only inexpensive, it also recycles a waste product that would otherwise end up in a landfill or sewer. There are many recipes for biodiesel available on the Internet, all of which include the following basic ingredients and equipment:

vegetable oil, 99+ percent methanol, lye catalyst (such as potassium hydroxide), a blender or mini-processor, a weigh scale, measuring beakers, translucent white plastic containers, funnels, duct tape and a thermometer. If you are mixing biodiesel with traditional petroleum diesel, you should be aware of any limits placed on the final mix ratio by the engine manufacturer. Many of them recommend the final mix be no more than B20.

At an international level, liquid biofuels provided about 2.7 percent of the world's road-transport fuel in 2010, with ethanol production reaching about 49 billion litres and diesel production reaching 19 billion litres.[3] Some countries are mandating that transport fuels have biofuel content. For example, the Canadian government requires an average renewable fuel content of 5 percent in gasoline.

Brazil and the United States are the world's leading producers of bioethanol. Thanks to its bioethanol industry, Brazil is no longer dependent on foreign oil. After OPEC's oil embargo of 1973, which sent the world's largest economies into a tailspin, Brazil began a national program to develop an alternative to gasoline using the nation's sugar-cane crops. Brazil now uses sugar cane to mass-produce bioethanol, which is blended with gasoline to provide fuel for vehicles. In 2010 sugar-cane bioethanol accounted for 41.5 percent of the country's light-duty transport fuel. It is estimated that more than 90 percent of new vehicles sold in Brazil are flexible-fuel vehicles (FFVs), which can run on pure bioethanol, pure gasoline or a mixture of the two. There is also a national mandate requiring all fuel sold in Brazil to contain at least 25 percent bioethanol. The country now enjoys energy independence due to the combined success of its domestic oil and bioethanol industries. The United States, the largest producer and exporter of bioethanol, primarily uses corn in the production process. In 2009 the country's 200 biorefineries produced a record-breaking 10.6 billion gallons of bioethanol.[4] Other parts of the world are producing bioethanol from various crops such as cereals and sugar beet in Europe, and corn and wheat in Canada, to offset fossil-fuel use. As part of its Action Plan 2000 on Climate Change, the Government of Canada established a national initiative to increase the supply and use of ethanol produced from biomass.

Worldwide commercial biodiesel production increased 7.5 percent in 2010, resulting in a five-year average (from the end of 2005 to 2010) growth rate of 38 percent per year.[5] Unlike bioethanol production, which is mostly concentrated in two countries, biodiesel production is spread across several countries. The European Union produced more than half of the world's supply, with Germany being the top biodiesel-producing country. Other significant producing countries are Brazil, Argentina, France and the United States. The highest growth in 2010 was seen in Brazil and Argentina, up 46 percent and 57 percent respectively, while the greatest fall in production happened in the United States, which was down 40 percent in the year. Some of the latest international large-scale biodiesel project announcements include soybean-based plants in Argentina, palm-based plants in Brazil and Malaysia, rapeseed-based plants in Germany, and plants in China that use waste cooking oil.

Benefits

By replacing fossil fuels such as oil, gasoline and petroleum diesel with liquid biofuels, society can significantly reduce greenhouse-gas emissions and reduce dependency on foreign oil. The production of biofuels can be relatively inexpensive, since it uses low-cost plants and organic waste as the energy source. Unlike fossil fuels, the crops used in the production of biofuels are renewable, since they can be replanted to provide an endless supply.

With the increasing popularity of liquid biofuels, new business opportunities are emerging in the farming industry for biomass crops to feed liquid-biofuel production processes.

Challenges

There are concerns that farmers may choose to grow crops for fuel instead of food, causing food shortages and higher prices. This is a legitimate fear that points to the importance of considering all trade-offs in how we use agricultural land. In some regions, farmers are expanding to grow both food and biofuel crops, resulting in economic prosperity and solving a fuel-shortage problem at the same time. One solution is to grow biofuel crops on land that is not suitable for food crops, and to use non-food crops such as switchgrass.

New Developments

First-generation biofuels, which are made from sugar, starch and vegetable oil, are now being replaced with second-generation biofuels that can be produced by using non-food crops such as switchgrass, or non-edible parts of food crops such as corn stems, leaves and husks. By avoiding the use of food crops, this new generation of biofuels is meant to overcome the potential problems of food shortages and price increases.

Bio-oil, also known as pyrolysis oil, can be created by subjecting biomass to pyrolysis, resulting in a thick, liquid, tarlike fuel. The process involves decomposition of the organic matter by heating it to a specific temperature in a closed container without oxygen. The output of this process is mainly char, which can be used as a solid biofuel, and bio-oil, which can be used to produce transportation fuels, although this development is still in the early stages of market entry.

A team of German and Chinese scientists have developed a new method for processing bio-oil directly into petroleum-type products. Having the whole process in one step is a breakthrough that simplifies production and increases energy efficiency; however, the process uses palladium, an expensive precious metal that scientists hope will be replaced someday with a cheaper alternative.[6]

In another development, algae are being used as biomass for producing bio-oil and are considered a third-generation biofuel feedstock. There are a few ways to extract oil from algae, but the simplest and most popular method is the oil press, which works like an olive press and extracts up to 75 percent of the oil from the algae. The algae can be harvested from waterways that have elevated nutrient levels due to fertilizer run-off, animal waste or industrial waste, or municipal waste-water treatment ponds that have elevated nutrients due to human waste. The algae absorb these nutrients as they grow; therefore, growing algae to produce bio-oil can have the beneficial side effect of removing nutrient contamination in waterways. Shallow water is ideal for growing algae because it allows more light saturation, leading to higher photosynthesis rates that provide more oxygen to promote the conversion of waste-water nutrients into algae. Many manufacturers of algae oil use a combination of mechanical pressing and chemical solvents to extract the oil.

GASEOUS BIOFUELS

There are two types of gaseous biofuels: biogas and syngas. Although they both can be used as a substitute for natural gas, biogas consists mainly of biomethane (CH_4), while syngas consists primarily of hydrogen (H_2) and carbon dioxide (CO_2).

BIOGAS

How It Works

Biogas, composed of roughly 60 percent biomethane and 40 percent carbon dioxide, refers to a gas produced by the biological breakdown of organic matter in the absence of oxygen.

Biogas is produced through anaerobic digestion (a natural process that occurs when biomass is subjected to an oxygen-free environment at a specific temperature). The resulting gaseous fuel can be used to generate electricity and heat in a number of ways: it can be burned in a boiler to produce high-pressure steam that drives a turbine; it can be injected into a combustion piston engine, which drives a shaft in a generator; or it can be used as a direct replacement for natural gas in providing hot water and space heating.

Real-Life Applications

Some farms use anaerobic digestion of cow manure to produce methane-rich biogas to generate electrical power, which they use or sell to the local power grid. Other renewable sources such as corn, cereal crops, sunflowers, Sudan grass, oil radish and other agricultural crops are also being used as biomass feedstock for the anaerobic digestion process.

In another agricultural example, wineries are creating biogas using grape by-products such as grape skins and seeds that otherwise would have been destined for a landfill. Inniskillin Winery, located in the Niagara region of Ontario, has partnered with a biogas developer and operator to create clean, renewable electricity from about 1,000 to 2,000 tonnes of grape pomace, which comprises grape skin and seeds. According to an Inniskillin representative, the biomethane gas that is produced by the decomposition of grape pomace will be captured and used to generate power for homes in the region.[7]

Many municipalities are installing facilities that capture biogas from their landfill sites to use as an energy source for their own operations or sell in order to generate revenue. Landfills are common sources of biomass used in the production of biogas. As organic materials naturally decompose, they produce and emit biogas that can be captured and injected into piston engines to drive generators that produce electricity for nearby industrial facilities and/or homes, or that can be fed into a local power grid. Alternatively, biogas can be cleaned and conditioned to raise its percentage of methane to the level required for injection into a natural-gas pipeline. Another way to obtain biogas from landfills is to separate organics from the rest of the garbage and put them into an anaerobic digester. Once all the organic material has been removed, a large portion of the leftover garbage (plastic, metal, etc.) can be recycled, thereby significantly reducing the amount of garbage going to the landfill.

Waste-water treatment plants can also carry out biogas production. In this case, human solid waste is separated from the liquid stream and put through the anaerobic digestion process by putting the solids in a temperature-controlled, oxygen-free, enclosed vessel. The resulting biogas is often injected into a piston combustion engine to generate electricity and heat for use by the waste-water facility.

Benefits

The biomass required for the production of biogas is widely available, can be replenished indefinitely, is carbon neutral and does not negatively affect future generations. Biogas production can provide an economic boost to farming communities or municipalities where much of the methane-producing biomass waste can be found and transformed into usable biogas.

The flexibility of biogas is a key benefit. As a gaseous fuel, it can directly displace natural gas and other types of fossil fuels to produce heat, steam, hot water and/or electricity, or be used as transportation fuel.

Challenges

For biogas systems to be feasible, the production costs on a per-unit-of-energy basis have to be relatively competitive with local natural-gas prices. Capital, operating and maintenance costs have to be considered. For example, capital costs may include building an anaerobic digestion facility. If the biogas is to be sold to a natural-gas utility company, there will be additional costs to condition the biogas to meet natural-gas quality levels and to build the infrastructure to deliver the biogas to the natural-gas pipeline.

New Developments

With massive amounts of organic materials currently going to waste, there is very good potential to tap into these unused resources and make a significant contribution toward solving global energy needs. Technologies are currently being developed and improved to capture or produce biogas and generate renewable energy from it in even more efficient and cost-effective ways.

For example, the biogas recovery rate from a landfill can be as low as 25 percent, but that can be improved significantly through the use of *bioreactors*, which tightly monitor and control the levels of moisture and air in the landfill to enhance biogas production.

Newly developed technology enables waste-water treatment plants to not only produce biogas from solid waste, but also extract and recycle phosphorous from liquid waste for use as organic fertilizer. The biogas can provide power for the fertilizer-production process, and leftover heat from the biogas process can be used to dry the fertilizer pellets before they are transported. This symbiotic relationship between bioenergy and fertilizer production provides economic, social and environmental benefits. It helps waste-water facilities get rid of unwanted phosphorous that clogs their pipes, thereby reducing maintenance and disposal costs, and it provides them with a revenue stream from selling the fertilizer. Phosphorous is a dwindling natural resource and an essential nutrient required to grow the world's food supply. Recycling phosphorous from waste water can continue indefinitely, unlike conventional mining, which will eventually deplete our natural phosphorous resources. Furthermore, conventional fertilizer production is very energy-intensive and emits a significant amount of greenhouse gases, so producing fertilizer from waste water with the help of biogas is much more energy-efficient and environmentally friendly.

SYNGAS

How It Works

Syngas is produced from the *gasification* of wood or other solid biomass. As a gaseous fuel, syngas is highly versatile, since it can be contained, stored and transported, and used as a direct substitute for natural gas to produce heat, steam, hot water and/or electricity. Gasification is a thermochemical process that differs from combustion because it uses just 20 to 30 percent of the oxygen needed for complete combustion. This "starved-air" process provides sufficient heat to chemically break down the biomass fuel into syngas, which is composed primarily of carbon monoxide, hydrogen and methane.

Real-Life Applications

The gasification process has been used to produce energy for over one hundred years; in earlier days, coal and peat, instead of wood, was fed into the plants. Today there are several industrial-scale and mid-sized gasification facilities around the world. As gasification systems become more efficient and cost-effective, it becomes increasingly feasible to shift from large, centralized facilities to smaller, decentralized facilities. We can expect to see more communities building small-scale, district energy facilities that use locally sourced wood waste instead of fossil fuel. An example can be found in the waterfront real-estate development Dockside Green, in Victoria, British Columbia, which feeds local wood waste into a small-scale gasification facility to generate heat and hot water for the buildings in the development.

Benefits

Suppliers of gasification systems claim they have lower air emissions and higher energy efficiencies than conventional combustion systems. Therefore, gasification systems are now being installed in residential areas, where they use local solid-waste biomass from tree trimmings and other types of waste wood suitable for the process.

Challenges

The gasification process is a complex multi-stage process that requires more investment in the facility setup and more upfront processing of the biomass fed into the system than a combustion plant. Once produced, the syngas must be conditioned before replacing natural gas to produce heat and electricity.

There is some public resistance to gasification facilities being installed in residential areas, due largely to the common misconception that gasification is just a fancy word for "incineration." Having said that, gasification plants are not completely free of pollution, so there are some valid concerns in this area that can be addressed through strict emission controls.

New Developments

The efficiency of gasification systems can be maximized with combined heat and power (CHP) systems whereby syngas is injected into a piston engine connected to a generator to produce both electricity and heat simultaneously. These systems are still in the demonstration stage. The University of British Columbia is installing a gasification CHP demonstration facility in Vancouver that will provide heat and power to the campus using waste wood.

Research and development is being carried out on systems to gasify other types of solid biomass, such as human biosolids extracted from waste-water treatment plants.

KEY CHARACTERISTICS OF BIOMASS ENERGY

Renewable? Yes, biomass is renewable because it can be replenished in nature indefinitely. For example, as long as trees are replanted and managed responsibly, there will be an endless supply of wood, and as long as humans and animals consume meat, vegetables, fruits, grains and dairy, there will be an endless supply of biosolid waste.

Sustainable? Yes, biomass is sustainable. Since biomass is renewable and carbon neutral, it poses no risk for future generations and is therefore a sustainable energy source as long as food crops are not negatively affected.

Provides Electricity? Yes, biomass is used to generate electricity either by generating heat that boils water to create high-pressure steam that turns the blades of a turbine, which drives an electric generator, or by producing a fuel that is injected into an engine that drives an electric generator.

Provides Heat? Yes, biomass is used to generate heat and/or hot water by combusting or gasifying solid biomass.

Delivers Constant Energy? Yes, biomass can be used to deliver continuous, non-fluctuating energy by stockpiling it and then feeding it into a biomass-energy process as needed.

Widely Available Supply? Yes, biomass is abundantly available in most regions around the world.

Easily Transported and Stored Supply? Yes, biomass can be physically transported by truck or rail and can be stockpiled with minimal maintenance requirements.

chapter nine

CLEAN-ENERGY TECHNOLOGIES

These technologies

play an important

SUPPORTING ROLE to

the green-energy sector.

THE BASICS

Unlike the alternative-energy technologies discussed in earlier chapters, these clean-energy technologies do not generate energy; rather, they make the delivery and/or storage of energy more clean or efficient and hence reduce our overall carbon emissions.

FUEL CELLS

A fuel cell is not an alternative-energy source but a highly efficient energy-conversion and delivery system, which typically uses hydrogen as its energy source. A fuel cell converts chemical energy from the reaction between hydrogen and oxygen into electricity through an electrochemical process. By doing so, it can very efficiently power a vehicle, electronic device or a remote power station, to name a few applications.

How It Works

There are several types of fuel cells, which all operate according to the same fundamental principles. Generally speaking, a fuel cell operates like a battery. One common type of fuel cell, known as proton-exchange membrane (PEM), has two electrodes, one positive (called the anode) and one negative (called the cathode), and an electron-blocking membrane between them, as shown in figure 9.1.

Hydrogen gas is passed over the anode and oxygen over the cathode. The hydrogen reacts to a catalyst coating on the anode that causes it to separate into negatively charged electrons (e^-) and positively charged ions (H^+). The positively charged hydrogen ions flow through the membrane toward the cathode. Since

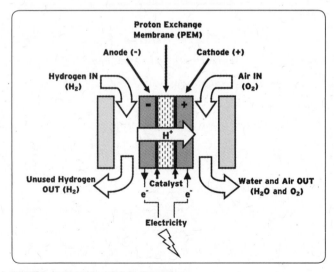

FIGURE 9.1 >>> A Typical Fuel-Cell Configuration

the membrane blocks electrons from flowing through it, they are forced to take a different path through an external circuit, where they are used for electrical energy on their way toward the positively charged hydrogen ions at the cathode. At the cathode, the electrons, hydrogen ions and oxygen combine to produce water (H_2O) through the chemical reaction shown in figure 9.2. As a result, electricity, water and heat are the only outputs of the fuel-cell process.

$$4H^+ + 4e^- + O_2^- \Rightarrow 2H_2O$$

Hydrogen   Oxygen   Water

FIGURE 9.2 »» Chemical Reaction: Fuel Cell

Benefits

Since fuel cells are chemically converting (instead of burning) hydrogen, there are none of the common by-products of combustion, such as greenhouse gases and other pollution. The only by-products are heat and water.

If the fuel cell's hydrogen comes from a renewable source such as biomethane, the entire end-to-end fuel process will be carbon neutral and will not contribute to a buildup of greenhouse gases in the atmosphere.

The energy efficiency of fuel cells is about 40 to 60 percent, which is better than the average energy efficiency of piston-combustion engines. Therefore, fuel-cell vehicles can travel farther than combustion-engine vehicles on the same amount of energy.

In the short term, hydrogen fuel cells are already being successfully used for stationary and portable power applications, and small specialized vehicle fleets with central fuelling depots, such as forklifts and municipal vehicles. Unlike batteries, fuel cells never run out of energy as long as they are being fed fuel (e.g., hydrogen and oxygen for a PEM fuel cell).

Challenges

For fuel-cell systems to be truly renewable, the hydrogen itself must come from a renewable source. Hydrogen is not available in nature in its pure form, so it must be extracted from other sources that contain hydrogen, typically hydrocarbons. The hydrocarbons come from various sources such as fossil fuels or biomethane.

If hydrogen for fuel cells is extracted from fossil fuels, then the end-to-end fuel-cell process is non-renewable and, even worse, contributes to an overabundance of carbon dioxide in the atmosphere. Therefore, when we consider the end-to-end energy chain involved with fuel cells, we have to think about where the hydrogen is coming from. Even though fuel cells themselves are emissions-free, if their energy source originates in hydrogen obtained from fossil fuels, they still contribute to greenhouse gases in the atmosphere. For example, fuel cells often use hydrogen extracted from natural gas, a fossil fuel that releases carbon dioxide in the chemical reaction to produce hydrogen, as shown in figure 9.3.

Another way to get hydrogen is by electrolyzing water to extract the hydrogen atoms, but this approach is truly renewable

$$CH_4 + 2H_2O \Rightarrow 4H_2 + CO_2$$

Natural Gas   Water   Hydrogen Carbon Dioxide

FIGURE 9.3 »» Chemical Reaction: Separating Hydrogen from Natural Gas

only if the electricity used for this process comes from a renewable-energy source such as solar, wind or hydro. Again, it's important to consider the entire end-to-end process to ensure that it does not depend on fossil fuels at any point in the process.

We've been hearing about fuel cells for a long time. Despite the billions of dollars that have already been invested in fuel-cell technologies, they are still a long way from being widely available for vehicle use on a mass scale. Some of the remaining technical, financial and infrastructure challenges include converting gasoline stations to hydrogen stations, finding sufficient quantities of renewable sources of hydrogen, transporting hydrogen to geographically dispersed filling stations, safely storing hydrogen on fast-moving vehicles, and achieving performance, driving distances, reliability and prices that are attractive to buyers.

CARBON CAPTURE AND STORAGE

Various types of carbon capture and storage (CCS) technologies are being developed to capture and store carbon dioxide emitted during the combustion of fossil fuels. Once captured, the carbon dioxide is usually stored underground. While some other clean-energy technologies focus on reducing the use of fossil fuels, CCS focuses on continuing the use of fossil fuels but reducing their carbon emissions into the atmosphere.

How It Works

The first step in the CCS process is to capture and separate carbon dioxide from other gases emitted during the combustion of fossil fuels. The three methods typically used are post-combustion, pre-combustion and oxyfuel combustion.

With the post-combustion method, carbon is captured from the gas coming off the power plant's exhaust after combustion. This method uses technology that is more mature than the others. A key advantage of this approach is that the technology can be installed at the location of the exhaust on existing or new fossil-fuel power plants; therefore, it can be retrofitted into hundreds of plants that are already operating.

In the pre-combustion method, the fossil fuel is partially oxidized before it is combusted to convert it into a gas containing mainly hydrogen and carbon dioxide. The hydrogen is then used for fuel, and the carbon dioxide is removed. This method cannot easily be retrofitted to existing plants because it requires a large degree of interaction with the process in the pre-combustion stage; however, it can be designed into new facilities.

The oxyfuel combustion method is very similar to post-combustion. The main difference is that the combustion is carried out with pure oxygen instead of air. As a result, the flue gas contains mainly carbon dioxide and water vapour, which can be easily separated. The main challenge with this method is that it is expensive to produce pure oxygen.

Once the carbon dioxide is captured through one of the three methods described above, it is compressed to reduce its volume so that it's easier and more efficient to transport, typically by pipelines or ships. Then it's pumped into depleted oil and gas reservoirs, saline formations or other underground geological formations where it's permanently stored.

Benefits

Once available, CCS will allow the continued use of fossil fuels for energy purposes with about 80 to 90 percent less carbon dioxide being released into the atmosphere in the process. CCS technology can significantly reduce carbon emissions from fossil fuels that contribute to global warming.

Another benefit is the potential for combining CCS with enhanced oil recovery, a process that involves injecting carbon dioxide into aging oil fields to boost recovery of crude oil by reducing the oil's viscosity, which makes it easier to transport it from deep in the ground up to a producing well. However, this is truly beneficial only if the recovered oil also undergoes CCS.

Challenges

CCS technologies consume a lot of energy. In particular, the carbon dioxide–capture stage is the greatest energy-consuming and most expensive part of the entire process. Therefore, research and development programs are concentrating their efforts on ways to improve the process of capturing carbon dioxide.

In addition to energy costs, CCS has other capital and operating costs that make the overall process relatively expensive. These costs include the installation of carbon-capture equipment, construction of a transport system, as well as compression, transportation and storage of the carbon dioxide. It has been estimated that adding a carbon-capture and storage system to a brand-new coal-power plant could add roughly 50 percent to the upfront capital cost.[1] In addition, retrofitting carbon capture and storage into existing power plants is more expensive than building the technology into new plants.

Although carbon-capture technologies are currently available in laboratories and small pilot plants, it will likely take several more years before they are widely used in fossil-fuel power plants and industries.

One of the world's first and largest demonstration CCS projects is the Boundary Dam Integrated Carbon Capture and Storage Demonstration Project in Saskatchewan. The project involves updating a coal-fired unit at the Boundary Dam Power Station into a producer of about 100 MW of clean electricity by retrofitting it with a carbon-capture system. The resulting captured carbon dioxide will be compressed and transported through pipelines and sold for enhanced oil recovery. This $1.24-billion project is to be financed through a partnership between the Government of Canada, the Government of Saskatchewan, SaskPower (an electric utility company) and private industry.[2]

SMART GRIDS

Smart-grid technologies enable utility companies, businesses and consumers to automatically monitor and optimize electricity usage, and detect and alleviate problems in real time in order to achieve energy-delivery systems with higher reliability, lower costs and less impact on the environment.

The "Smart Grid" is also a term used to describe a long-term vision to overhaul nationwide power grids by adding electronic intelligence that will make them capable of automatically routing power in optimal ways, responding to a wide range of conditions and seamlessly integrating alternative-energy sources into the grid.

The smart-grid technologies described below are already being used for buildings, communities and municipalities and may also become components of the nationwide Smart Grid in the future. Examples of smart-grid technologies include smart metering and phasor measurement units.

How It Works

The power grid transports electricity from power-generating facilities to consumers through a network of electrical transmission lines and is typically owned and maintained by utility companies. Smart-grid technologies encompass a range of applications that enable utility companies, businesses and residential consumers to automatically monitor, control and optimize energy delivery and usage along this network and into buildings.

For example, smart meters are advanced, digital meters installed in homes and commercial buildings that allow utility companies to receive automated, real-time remote readings of their electricity consumption. These smart meters track electricity usage levels and identify what time of day the electricity was consumed with more detail than regular meters. Some smart meters can even track electricity usage patterns on individual appliances or electronic devices. In some cases, smart meters enable consumers to preset times when certain appliances or devices (such as a phone charger) will turn on and off in order to avoid running them during times of peak demand when energy prices are highest.

Another smart-grid technology, a phasor measurement unit, enables utility companies to monitor power quality by measuring the waveforms of electricity on a power grid. In some cases, it triggers an automatic response to any problems that are detected.

Benefits

By making power grids operate more efficiently and providing utilities and consumers with the ability to manage energy consumption on a more detailed level, smart-grid technologies help reduce energy costs and usage, thereby minimizing greenhouse-gas emissions.

With applications that rapidly detect and correct problems in the power grid, these technologies enable "self-healing," which means that grids can be programmed to automatically repair themselves or switch to a different power routing when something goes wrong, meaning fewer service disruptions.

The total demand on a power grid varies significantly over time and can overload during times of peak demand. Smart-grid technology helps smooth out the peaks and valleys by enabling utilities to track consumer usage on a real-time basis and offer consumers price incentives to reduce their demand at peak times. As a result, the power load on the grid is stabilized, outages are avoided and money is saved. Furthermore, by smoothing out demand levels, utilities can avoid building new power capacity to handle extremely high peak levels.

Challenges

Smart-grid technology is being resisted by some who view it as a way for governments to track and even control consumers' energy consumption in their own homes, thereby invading their privacy.

While most smart-grid technologies enable consumers to save money by avoiding peak-time energy prices, the implementation costs of those systems can erase some of the savings. Furthermore, in some cases, smart technologies require infrastructure changes that need approval by government authorities, resulting in additional upfront capital expenditures.

ENERGY STORAGE

The main disadvantage of many types of green-energy sources such as solar, wind and ocean is that they produce intermittent electricity, whereas fossil fuels provide consistent electricity on demand. This is a big problem for utility companies, which need constant, reliable power sources to meet consumers' energy needs, and is also the reason that some utility companies charge financial penalties to green-energy producers who fail to deliver certain power-supply targets. In some cases, utilities refuse to work with green-energy providers altogether.

Imagine there was a way to easily and cost-effectively store excess electricity during peak production times—when the wind is blowing or the sun is shining—and save it until it is needed to create a more reliable, constant flow of electricity. Energy-storage technologies are being developed that will do just that.

How It Works

There are several types of energy-storage technologies being explored by researchers around the world; however, each of them has pitfalls that need to be overcome. A few of those technologies are explained below.

As mentioned in chapter six, pumped-hydro storage is currently the most widely used type of energy-storage technology. This process involves developing two large water reservoirs at different elevations. Excess electricity (e.g., from a wind turbine turning at night when demand is low) is used to pump water from the lower reservoir to the higher one. When electricity is required, the water in the higher reservoir is released through turbines to generate power. The main disadvantage of this system is that it is expensive to build.

Advanced battery systems can also be used for energy storage. For example, *flow batteries*, which store and release energy through a reversible electrochemical reaction between two electrolytes, are already being used to smooth out fluctuations in wind-energy supply by storing excess electricity when the wind is blowing for later use when there is no wind. However, batteries lose some of their ability to hold a charge over time so they need to be periodically replaced, thereby increasing operating costs.

A flywheel energy-storage system works by drawing excess electricity from a primary energy source (e.g., a wind turbine) during peak production to spin a heavy cylinder (flywheel) at very high speeds, thereby maintaining the energy in the system as kinetic energy. When the primary energy source either decreases or stops production (e.g., the wind stops blowing), the power stored by the rotating flywheel is converted to electrical energy through an electric generator to supply power to the grid as needed. A significant disadvantage of flywheels is that they have the potential risk of breaking apart if they exceed certain speeds of rotation. Also, flywheels do not store energy for very long; some can lose 20 to 50 percent of their energy in a matter of hours.

Hydrogen is also being used as a medium to store excess electrical energy. For example, surplus electricity from a green-energy source such as a wind turbine can be temporarily converted to hydrogen through electrolysis, a process by which an electric current causes water to separate into its components, hydrogen and oxygen, in a chemical reaction, as shown in figure 9.4.

$$2 \; H_2O \Rightarrow 2 \; H_2 + O_2$$

Water   Hydrogen Oxygen

FIGURE 9.4 >>> Chemical Reaction: Electrolysis of Water

The hydrogen is then compressed and stored. When the electricity is needed, the stored hydrogen is converted back to electrical energy by injecting it into fuel cells. One drawback is that the overall efficiency of using hydrogen as a storage medium is typically lower than other methods of energy storage.

Benefits

Overcoming the issue of energy storage would certainly accelerate the adoption of green energy as a replacement for fossil fuel–based energy. Such a shift would lead to lower greenhouse-gas emissions, greater energy security and lower risk of oil spills and coal-mine disasters.

Challenges

There are challenges with all of the current storage technologies being explored.

The financial feasibility of currently available energy-storage technologies varies widely and in many cases prevents implementation.

Some energy-storage technologies may be harmful to the environment. For example, batteries contain toxic materials that are difficult to dispose of, and pumped-hydro reservoirs may cause flooding.

The green-energy industry is still waiting for a major break-through in energy storage that will provide a cost-effective, environmentally friendly solution leading to significantly wider adoption of green energy.

chapter ten

HOUSEHOLD ENERGY CHOICES

This chapter explores

household energy choices

and provides information

to help individuals

make **INFORMED ENERGY**

CHOICES in their

everyday lives.

REDUCING OUR CARBON FOOTPRINT THROUGH GREEN ENERGY

As green energy enters the mainstream, we are being offered more energy choices in our day-to-day lives. Renewable energy is now being implemented in residential developments as more developers adopt greener energy options such as geoexchange heating and waste-to-energy facilities for district power and heating. Meanwhile, our local car dealers are offering us electric, hybrid and highly fuel-efficient vehicles.

As a result, most of us are starting to face decisions that will have an impact on our households and lifestyles. What kind of energy do we want to use in our homes and vehicles? What are the options? What are the pros and cons?

I am often asked what is meant by "carbon footprint." This term refers to the total amount of carbon emitted into the atmosphere as a direct result of your personal day-to-day activities. While it encompasses every aspect of your life, from eating to shopping and recreation, the two factors that contribute most to your personal carbon footprint are your household energy use and personal transportation. Because fossil fuels are the greatest contributors to human-related carbon emissions, the most effective way to significantly reduce your carbon footprint is to decrease or eliminate the amount of fossil fuel used for your household power and heating, and vehicles.

ENERGY CONSERVATION IN THE HOME

Before exploring green-energy options for a household, the first step is to evaluate existing energy usage and look for ways to conserve. This does not mean that homeowners have to sacrifice comfort. Conservation starts with simply looking at ways to decrease energy waste in the home and eliminate some of the costs and environmental impacts of energy usage without giving up the luxuries that power and heat can provide. After reducing waste, we will have a better understanding of how much energy we actually need to run the household.

The first step is to check utility bills to determine annual energy consumption. Then, evaluate where and when most of the energy is being used to see where there may be opportunities to conserve energy. If possible, hire a professional residential-energy auditor to provide advice on the most effective and economical ways to upgrade your home's energy performance.

Some of the greatest reductions in energy consumption can be achieved by taking the following actions:

>>> develop personal habits to reduce household energy waste such as turning off lights when they are not needed

>>> replace old appliances with newer, energy-efficient models

>>> add insulation to the attic to minimize heat loss

>>> replace old windows with better-insulated ones

>>> install thermostats that can be preset to automatically turn the heat down when you are routinely out of the house

>>> use energy-efficient electronics and lighting such as LED lights, and turn them off when they are not required

Homeowners have reported significant reductions in their annual energy consumption through these energy-conservation methods. As a result, they are in a better position to select an appropriately sized green-energy system to handle the remaining energy requirements of the household. By reducing energy waste in the home before choosing a new green-energy system, you will save money by installing a smaller, less expensive system.

GREEN-ENERGY OPTIONS FOR THE HOME

Take a closer look at where the power and heat in your home are coming from. If they are based on fossil fuels, then you might want to consider installing a green-energy system to provide all or some of your energy needs and thus reduce your carbon footprint. Before choosing a green-energy system for your home, gather additional information from specialists who sell and install these systems. If possible, tour homes that are already using them and talk to homeowners who have experience with them.

ELECTRICITY

First, find out where the utility company is getting the electricity used in the home. If it is produced by hydroelectric dams and/or wind farms, then it is already coming from a renewable-energy source. But if the electricity is produced by any kind of fossil-fuel power plant, then explore green-energy systems for the home such as a residential solar- or wind-power system. Installers of these systems will help you determine if you have a sufficient amount of sunshine or wind to make such a system viable.

SPACE HEATING

Next, find out how the house is being heated. For example, if heat comes from electric-baseboard heating, then the home's heat and electricity are both from the same energy source. If the house has forced-air heating, the heat is likely coming from a furnace, so determine what kind of fuel is used in the furnace. If the house has radiant heating, all the heat is coming from the hot-water system, so the energy source for that system must be identified.

If the source of energy used for heating the house is a fossil fuel such as natural gas, oil or propane, then consider switching to a green-energy system for space heating such as geoexchange or solar heating.

HOT WATER

In most homes, hot water for domestic use such as showers, dishwashers and laundry comes from hot-water tanks that are heated by natural gas, oil or propane, all of which are fossil fuels. Sometimes, hot water comes from electric hot-water tanks that may derive their electricity from either a fossil fuel, like coal, or a renewable energy source, like hydro. If the hot water is being heated by a fossil fuel, then changing the hot-water system to one heated by green energy will reduce your carbon footprint. Usually a home's energy source for space heating is the same as for hot water, so a green-energy system such as geoexchange or solar heating would typically be installed to provide heat for both space heating and hot water.

NEW ENERGY OPTIONS FROM UTILITY COMPANIES

Some utility companies are now offering consumers the option to voluntarily pay a premium to receive green energy instead of fossil fuel–based energy. For example, a utility company in British Columbia, FortisBC, is now offering some of its customers the option to buy biomethane generated from biomass, instead of natural gas. Consumers who opt to pay the higher price for this biomethane receive a carbon-tax reduction from the provincial government to offset the extra cost.

In many other parts of the world, especially in Europe, utility companies now offer their customers the option to buy electricity and heating from green-energy sources such as wind, biomass and others.

LEED CERTIFIED RESIDENTIAL DEVELOPMENT

If you are buying a new home, there are now residential developments that meet Leadership in Energy and Environmental Design (LEED) certification. This certification system recognizes building projects that have implemented sustainable building practices and is an internationally accepted benchmark for the design, construction and operation of high-performance green buildings. It rates performance in five categories: sustainable site development, water efficiency, energy efficiency, materials selection and indoor environmental quality. LEED certification is awarded on a 100-point system, whereby some points are awarded for implementing renewable-energy systems and energy conservation methods that reduce greenhouse-gas emissions. Developments can achieve platinum, gold, silver or certified levels, depending on how many points they earn.

For example, Vancouver's new convention centre, which opened in 2009, achieved Platinum certification, the highest LEED rating. To date, it is the only convention centre in the world to receive this level of LEED certification.

PERSONAL VEHICLES

Nowadays, consumers have a vast array of vehicle types to choose from such as high fuel efficiency, flexible-fuel, electric and hybrid vehicles. Fuel-cell vehicles may also be available in the future.

HIGH-EFFICIENCY

High-efficiency vehicles use traditional combustion engines but are designed to be very energy-efficient, going farther on a tank of gas than other combustion vehicles. These efficiencies are achieved through designing smaller and more aerodynamic vehicles that are built with lighter-weight materials. But although they use less fuel than the average vehicle, they still rely on fossil fuels, which emit greenhouse gases during the combustion process.

FLEXIBLE-FUEL

There are already millions of flexible-fuel vehicles (FFVs) on the road today, including more than half a million in Canada and more than 8 million in the United States alone.[1] These vehicles have internal-combustion engines designed to run on more than one fuel, including gasoline combined with either ethanol or methanol fuel, which are stored in the same tank.

FFVs are most commonly found in Brazil, Europe, Canada and the United States and usually run on a blend of 15 percent gasoline with 85 percent ethanol, known as E85 fuel.

DIESEL, PROPANE AND NATURAL GAS

Some vehicles use internal-combustion engines but replace gasoline with other fossil fuels including diesel, propane or natural gas, which are slightly more energy-efficient.

Diesel and propane vehicles consume less fuel than gasoline-powered vehicles travelling over the same distance, thereby providing some cost savings to their owners. Although they emit slightly less carbon than gasoline-powered vehicles, they still use fossil fuels that contribute to an overabundance of carbon in the atmosphere, so they do not qualify as green-energy solutions.

Natural gas is a fossil fuel composed mainly of methane (CH_4). Natural-gas vehicles are slightly more fuel-efficient than those powered by gasoline, and natural gas is cheaper than gasoline in many locations. Natural gas emits less carbon than other fossil-fuel derivatives, including gasoline. Furthermore, there are still large, untapped natural-gas resources available in North America, so there is no need to import it from overseas. Some individuals have safety concerns about carrying compressed natural gas on board vehicles; however, manufacturers are quick to point out that it is no more dangerous than carrying a tank of gasoline. One of the key challenges is that there are still relatively few natural-gas vehicles available for sale, and not many gas stations carry natural-gas fuel. From an environmental point of view, although natural gas emits less carbon than other fossil fuels, it is still a fossil fuel that contributes significantly to an overabundance of carbon in the atmosphere.

If a driver used biomethane to fuel a natural-gas vehicle, that would be a truly renewable solution. Biomethane has the same chemical makeup (CH_4) as natural gas but is produced from biomass. When biomethane is extracted from biomass and combusted to produce energy, it emits the carbon that would have been released by the biomass anyway (as it died and decayed through the natural carbon cycle). This is in sharp contrast to natural gas, which is extracted from underground and combusted to emit "new" carbon into the atmosphere that would otherwise stay locked in the ground indefinitely. Therefore, natural-gas vehicles can only be truly green if they use biomethane instead of natural gas.

ELECTRIC

Electric vehicles use electric motors and batteries instead of traditional combustion engines. They plug into electrical outlets to charge the batteries and do not use any gasoline, diesel or propane; therefore, assuming their electricity source is renewable, they can be 100 percent free of greenhouse-gas emissions.

Even if their electrical source is based on fossil-fuel power such as coal, when we consider the end-to-end process including building and fuelling them, electric vehicles can still be more efficient than traditional combustion-engine vehicles. Electrical vehicles are quiet and relatively simple to maintain. Depending on local electricity prices and the cost of replacement batteries, they can also be relatively inexpensive to operate.

However, electric vehicles also have a few disadvantages. Currently the upfront price for electric vehicles and their batteries is relatively high. For instance, the Nissan Leaf electric car debuted in the United States in 2010 with a manufacturer's suggested retail price (MSRP) of $32,780 and in Canada in 2011 with a MRSP of $38,395. Electric cars, including the Leaf, have limits on the driving distance between charges (usually up to 100 miles or 160 kilometres), so owners have to develop personal routines for when and where to charge them on a regular basis. Furthermore, drivers have to allow enough time for the batteries to be charged between uses. The electric vehicle will greatly reduce the carbon footprint of car owners who do a lot of city commuting and have access to renewable electricity such as hydro. If that electricity happens to be inexpensive, that's even better, as it makes the vehicles both environmentally friendly and economical to operate.

HYBRID

Hybrid vehicles use technology that typically captures kinetic energy when the brakes are applied and converts it into electricity, which is stored in batteries to power the vehicle. When the vehicle is travelling at lower speeds or idling at stoplights, it runs silently on this self-generated electricity. When the vehicle reaches higher speeds, the combustion engine takes over and it runs like a regular vehicle, but with a boost from the battery whenever extra speed is needed. This combination of electric- and fuel-based power significantly reduces the overall amount of fuel needed to operate the vehicle and can save more than 50 percent on fuelling costs. The upside of hybrid vehicles is that they provide many of the benefits of electric vehicles, including very high efficiency and a quiet ride, without being dependent on external charging of the battery.

FUEL CELL

Fuel-cell vehicles could someday offer a truly emissions-free solution to personal transportation. They run on hydrogen instead of gasoline, and the hydrogen can come from a large variety of renewable sources such as biomethane or electrolysis of water.

There are a few limited-production fuel-cell vehicles being offered today. Honda offers leases on a few hundred fuel-cell cars in southern California, where there are some designated hydrogen-fuelling stations for customers to use. Mercedes has also leased out about 200 fuel-cell vehicles in selected European markets, and they recently introduced their lease program into California. Both manufacturers' programs sold out immediately and have customers on waiting lists.

Fuel-cell vehicles are still in the development stages and face significant obstacles such as access to hydrogen-fuelling stations and the need for a large supply of hydrogen from renewable sources to make them truly emissions-free. The ultimate vision for a sustainable fuel-cell vehicle is one that could be refilled at either a home fuelling station or local public fuelling stations, using hydrogen from a purely renewable energy source.

chapter eleven

COMPARING ALTERNATIVE-ENERGY TYPES

This chapter

SUMMARIZES AND

COMPARES the various

types of alternative

energy described

in earlier chapters.

KEY CHARACTERISTICS

We can now draw some comparisons and conclusions about various types of alternative energy including solar, wind, earth, hydro, ocean and biomass, all of which are alternatives to fossil fuel–based energy sources. These options can be compared based on the following key characteristics, which are also summarized in table 11.1.

RENEWABLE?

The alternative-energy sources examined in this book are renewable, since their energy sources are replenished in nature and therefore can continue indefinitely. Solar, wind, hydro, wave, biomass and geoexchange systems all derive their energy from the sun. Geothermal energy originates from the earth's core, and tidal energy comes from the gravitational forces of the moon and sun. Since the sun, moon and earth provide an endless, abundant source of energy, these renewable-energy options have the potential to meet all the energy needs of the world's population on an ongoing basis.

SUSTAINABLE?

The alternative-energy sources examined in this book are sustainable because they can be used by the current generation with minimal negative effect on the economic, social or environmental needs of future generations. Biomass energy emits some pollution in the combustion process that could potentially affect future generations; however, current technology is very effective at minimizing the harmful emissions from this process. Green-energy technologies often have some form of fossil fuel–based energy consumed during their end-to-end process, such as the energy that goes into the making of steel for turbines. Therefore, green energy does have some negative impact on future generations through that process, but this impact is minimal compared to that of fossil fuels.

PROVIDES ELECTRICITY AND HEAT?

Electricity is directly produced from every form of alternative energy except geoexchange, which is used specifically for space heating and cooling in buildings. Solar and biomass energy can be used to *directly* produce both electricity and heat. Geothermal is most often used to produce electricity, although it can be used to directly provide heat to nearby buildings. Other energy types such as wind, hydro and ocean can only produce heat *indirectly* by generating electricity that is then used for electric heaters.

DELIVERS CONSTANT ENERGY?

Some alternative-energy types, including earth, biomass and hydro, deliver a constant flow of energy. Others fluctuate, such as solar, wind and ocean energy, thereby creating challenges for utility companies that require consistent, reliable energy. To smooth out these fluctuations, technologies are being developed to store excess energy during peak production times and then release the energy when it is needed.

WIDELY AVAILABLE SUPPLY?

Generally speaking, most alternative-energy sources, including sun, wind, oceans, rivers and biomass, are available in most regions around the world for human use. Of course, some areas have more than others, but from a global point of view there are plenty of these natural resources all around us. The main exception is geothermal energy, which is only accessible in certain locations, typically away from populated areas.

EASILY TRANSPORTED AND STORED SUPPLY?

The transportation and storage of energy sources are of critical importance when assessing any type of alternative energy. Hydro and biomass energy have sources that can be easily stored and transported on a commercial scale. Water can easily be stored in reservoirs and transported through tunnels or pipes. Biomass organic matter such as wood can be physically stockpiled and transported by trucks, trains or ships.

However, the sources of solar, wind, ocean and earth energy cannot easily be transported or stored. Solar systems only produce energy when the sun is shining, and it is impossible to store and transport sunshine. Wind turbines only produce power when the wind blows, and it is also impossible to store and transport wind. Ocean-energy systems only produce power when waves are rolling or tidal currents are rushing through narrow passages, and these cannot be stored or transported. Geothermal sources such as hot-water springs or geysers lose their heat energy if stored or transported (and insulated

ENERGY TYPE	Renewable?	Sustainable?	Used for Electricity?	Used for Heat?	Delivers Constant Energy?	Widely Available Supply?	Easily Stored and Transported?
Solar	✓	✓	✓	✓		✓	
Wind	✓	✓	✓			✓	
Geothermal	✓	✓	✓	✓	✓		
Geo-exchange	✓	✓		✓	✓	✓	
Hydro	✓	✓	✓		✓	✓	✓
Ocean	✓	✓	✓			✓	
Biomass	✓	✓	✓	✓	✓	✓	✓

TABLE 11.1 ⇒ Key Characteristics of Alternative-Energy Types

pipes that would maintain more heat are expensive), so power plants are constructed at the geothermal sites. In contrast, geo-exchange systems use heat stored in the ground that can only be transported short distances, so these systems are installed very near to the buildings they heat.

COSTS

Arguably, the greatest barriers to implementing any kind of alternative energy are the costs, including both upfront capital costs and ongoing operating costs.

Alternative-energy costs depend on many factors such as resource availability, subsidies, policy incentives, system sizes and locations. Table 11.2 shows typical energy-production costs for large-scale installations as reported in the *Renewables 2011 Global Status Report*.[1] These costs, which exclude subsidies and incentives, assume good locations and resource availability. Optimal conditions could yield lower costs and less favourable conditions could result in higher costs.

Solar-PV electricity is clearly still the most expensive type of alternative energy, due to its relatively low energy-efficiency (typically 15 to 20 percent) and the high manufacturing costs of solar-PV panels at this time. The relatively high costs for residential and commercial solar-PV systems are shown in tables 11.2 and 11.3.

ENERGY TYPE	SIZE OF FACILITY	TYPICAL ENERGY COSTS (US cents/kWh)
Solar-PV Electricity	Up to 100 MW Utility-scale	15 - 30
Solar Heating	Large/ District Size	1 - 8
Wind	Onshore, Large Turbine Size	5 - 9
Geothermal	Up to 100 MW Large Plant Size	4 - 7
Geoexchange	Up to 10 MW District Plant Size	0.5 - 2
Hydro	Up to 18,000 MW X-Large Plant Size	3 - 5
Biomass Electricity	Up to 20 MW Medium Plant Size	5 - 12
Biomass Heat	Up to 20 MW Medium Plant Size	1 - 6

TABLE 11.2 >>> Typical Production Costs of Alternative-Energy Types (Large-Scale)

ENERGY TYPE	SIZE OF FACILITY	TYPICAL ENERGY COSTS (US cents/kWh)
Solar-PV Electricity	20 - 100 MW Solar Home System	40 - 60
Solar Hot-Water/ Space Heating	2 - 5 m^2 District Size	2 - 20
Wind	0.1 - 3 kW Household System	15 - 35
Pico-Hydro	0.1 - 1 kW	20 - 40
Mini-Hydro	100 - 1,000 kW	5 - 12

TABLE 11.3 >>> Typical Production Costs of Alternative-Energy Types (Residential)

On the other hand, the costs of solar-heating systems are among the lowest, since they have energy efficiencies of 50 to 80 percent, meaning that 50 to 80 percent of the radiant energy striking the solar collectors is converted into heat energy for the building. For instance, a solar hot-water/space heating system is one of the least expensive types of green energy that you can install in your household, as shown in table 11.3.

Wind-power costs are moderate compared to other types of alternative energy, but are still higher than traditional fossil-fuel costs. Turbine component costs have fallen over recent years due to increasing production volumes, and turbine technology has improved in power efficiency and reliability. However, the level of production for turbine components still has not reached sufficient volume to achieve the economies of scale needed to compete with fossil-fuel plants. Also, turbine towers and blades are difficult and costly to transport, and installation requires expensive equipment and highly specialized and experienced labour. There are also substantial costs involved in building infrastructure to transmit the electricity from remotely located turbines to the power grid.

The cost of geothermal energy is generally lower than that of wind, but still higher than fossil-fuel power plants. Since geothermal hot water and steam are not portable without losing heat energy, a geothermal plant must be built near the site of the resource. Therefore, there are substantial upfront costs involved in constructing the power plant in a remote area, creating an access road and then building transmission lines from the plant to the power grid. Furthermore, drilling costs involved in geo-thermal projects can be very expensive.

Geoexchange is a relatively affordable type of alternative energy, since the systems can use a small amount of heat found in the ground or a body of water to heat a whole building. This is because geoexchange systems have an energy efficiency of about 400 percent or even higher. The largest cost in geoexchange systems is usually drilling and excavation.

Hydro energy typically has low, stable costs, since it uses free-flowing water and conventional equipment. Hydro systems have been around for a long time, so their technology is well established and operates very reliably once installed, with minimal need for operators. Hydropower plants also have very long operating lives, which contributes to their overall low cost per kWh.

Biomass energy is one of the most feasible and profitable types of alternative-energy options, since it is fuelled by widely available, low-cost biomass such as human or animal waste, trees and plants, instead of fossil fuels.

FINAL THOUGHTS

Oil spills, coal-mine accidents, climate change, dependency on foreign oil, dwindling oil reserves and rising fuel prices are just a few reasons why many of us are seeking alternatives to fossil-fuel energy. As a result, we are starting to consider new energy options for our households and day-to-day lifestyles, ranging from the type of vehicles we drive to the kinds of electricity we use in our homes and businesses.

At present about 80 percent of the world's energy consumption comes from fossil fuels. Meanwhile, we are in a difficult situation where developing countries are increasing fossil-fuel consumption to bring transportation and electricity to millions of poverty-stricken people, and developed countries refuse to reduce their consumption until developing countries are willing to do the same—so it's a stalemate.

But there are solutions. A major new report by the United Nations–supported Intergovernmental Panel on Climate Change (IPCC) released in June 2011 underscores the incredible environmental and social advantages of a future powered by renewable energy. The report concludes that we could meet as much as 43 percent of global energy demand with renewable sources by 2030, and up to 77 percent by 2050,[2] largely using biomass, wind and solar energy. According to the report, renewable-energy types are already developing rapidly: of the 300 GW of new electricity-generating power plants added globally from 2008 to 2009, nearly half (140 GW) came from renewable sources.[3] Furthermore, the report confirms that global renewable-energy resources are substantially greater than both current and projected future global energy demand. In other words, we really don't need fossil fuels to meet our energy needs.

On a more local level, I hope this book helps readers to identify and select green-energy options so we can make these changes in our homes and lifestyles sooner rather than later. I also hope it encourages individuals, businesses and governments to incorporate sustainability into all areas of our lives to ensure that we are able to meet our short-term and long-term financial, social and environmental needs as well as those of future generations.

chapter twelve

FREQUENTLY ASKED QUESTIONS

This chapter provides

short **ANSWERS** to

frequently asked

questions about

green energy.

WHAT IS GREEN ENERGY?

Green energy is a common term for renewable energy.

WHAT IS RENEWABLE ENERGY?

Renewable energy refers to energy types that are naturally replenished and continue indefinitely. These include solar, wind earth, hydro, ocean and biomass.

WHAT IS ALTERNATIVE ENERGY?

Alternative energy refers to energy types that are alternatives to fossil fuels, such as solar, wind, earth, hydro, ocean, biomass and nuclear energy. The only difference between alternative energy and renewable energy is that alternative energy includes nuclear power, which is an alternative to fossil-fuel energy but is not renewable. For this reason, and because of the risks it poses, nuclear energy is not discussed in this book.

WHAT IS SUSTAINABILITY?

Sustainability refers to the ability to meet current financial, social and environmental needs without harming or compromising future generations' ability to meet those same needs.

WHAT IS THE DIFFERENCE BETWEEN NET METERING AND FEED-IN TARIFFS?

A feed-in tariff is a government-imposed pricing structure that utility companies are required to pay to end users for renewable electricity that those users generate and feed into the grid. Net metering, offered by some utility companies, allows consumers with small renewable-energy systems to feed their unused electricity back into the power grid, causing their electrical meters to run in reverse. This reduces their electricity bills and can earn them credit toward future bills.

WHAT IS THE DIFFERENCE BETWEEN A CARBON TAX AND CAP-AND-TRADE PROGRAM?

A carbon tax is placed on fossil-fuel energy purchased for heat, electricity and transport fuel in order to encourage both businesses and individuals to reduce their fossil-fuel consumption. Another method involves taxing utility companies on their purchases of fossil-fuel energy. Alternatively, companies may be taxed on the amount of carbon they emit during day-to-day operations. Carbon-tax programs are known as *revenue-neutral* if the government returns 100 percent of the revenue it collects from carbon tax to taxpayers through tax reductions in other areas.

A cap-and-trade program typically imposes a cap on the number of tonnes of carbon dioxide that are allowed to be emitted by an entity, such as a country or industry. Those that keep their emissions below the cap can sell their remaining allowance, converted from tonnes to carbon credits, on a carbon market, while those that exceed the cap must buy credits or face penalties.

WHAT IS THE DIFFERENCE BETWEEN A CARBON CREDIT AND CARBON OFFSET?

A carbon credit is a financial instrument that is transferable and saleable. One carbon credit represents one tonne of carbon dioxide emitted by the burning of fossil fuels. Carbon credits provide businesses or countries with a financial benefit for reducing carbon-dioxide emissions.

Carbon offsets are more like donations than financial instruments, and they are not transferable or saleable. Individuals or businesses can voluntarily offset a portion of their carbon footprint by paying for carbon offsets through brokers who invest that money in green projects that are certified to reduce carbon-dioxide emissions.

WHAT IS THE DIFFERENCE BETWEEN SOLAR PANELS FOR ELECTRICITY AND FOR HEAT?

For solar electricity, it is common to use rooftop solar panels containing PV cells, which include chemically treated silicon material.

When struck by sunlight, this material sets electrons free to create an electrical current. Solar heating uses rooftop panels known as solar collectors to capture the sun's radiant energy and convert it into thermal energy for hot water and/or space heating. As water passes through the collectors, it is heated by the sun.

WHERE DOES WIND ENERGY COME FROM?

Wind is caused by differences in air pressure due to the uneven heating of the earth's surface by the sun. The earth's surface absorbs sunlight and re-radiates it back out in the form of heat. Some parts of the earth absorb more solar energy than others depending on various factors such as the angle of the sun's ray's reaching the earth's surface and how different surfaces absorb or deflect the sun's rays. Land areas usually absorb more solar energy than bodies of water, so the air over land usually gets warmer than air over water. As air warms, it expands and rises, and nearby cooler, denser air rushes in to take its place, creating wind.

WHAT IS THE DIFFERENCE BETWEEN GEOTHERMAL AND GEOEXCHANGE ENERGY?

Geoexchange (also known as ground-source heating) and geo-thermal systems are similar, as they both derive their energy from underground. However, geothermal energy originates from heat (thermal) energy at the earth's core, while geoexchange energy originates from solar radiation absorbed by the ground.

WHERE DOES HYDRO ENERGY COME FROM?

Hydro energy originates from the sun, which causes water in oceans and lakes to evaporate and form clouds. Subsequently the vapour in the clouds condenses back into water and falls to the ground as snow or rain, which often flows into rivers that can be used to generate hydroelectricity.

WHAT IS THE DIFFERENCE BETWEEN OCEAN-TIDE ENERGY AND OCEAN-WAVE ENERGY?

Although tidal and wave energies are both related to ocean waters, they are traced to different sources. Tidal-energy technologies capture the kinetic and potential energy contained in moving water caused by tides. Wave-energy technologies typically use floating offshore mechanical devices that extract potential and kinetic energy from the up-and-down motion of waves.

HOW IS BIOMASS CARBON NEUTRAL?

Biomass material is derived from living or recently living organisms such as trees, plants, animals, vegetables and grains. These biological organisms absorb carbon dioxide as they grow and release it when they die and decay, or burn, as part of the natural carbon cycle. Because biomass releases roughly the same amount of carbon dioxide as it absorbs over its lifetime, it is carbon neutral.

ARE FUEL CELLS A TYPE OF GREEN ENERGY?

A fuel cell is not a green-energy source but a highly efficient energy-conversion and delivery system that typically uses hydrogen as its energy source. A fuel cell converts chemical energy from the reaction between hydrogen and oxygen into electricity through an electrochemical process. However, in order for the entire end-to-end fuel process to avoid contributing to a buildup of greenhouse gases in the atmosphere, the fuel cell's hydrogen must comes from a renewable source such as biomethane.

NOTES

CHAPTER ONE **WHY GREEN ENERGY?**

1   REN21. 2011, *Renewables 2011 Global Status Report*, 17.
2   IPCC, "Energy Supply," in *Climate Change 2007: Mitigation,* 261.
3   Dr. Colin J. Campbell, email message to author.
4   United States Energy Information Administration (EIA) statistics cited in Natural Gas Supply Association, "Natural Gas Supply."
5   Kanter. "Carbon Trading: Where Greed is Green."

CHAPTER THREE **SOLAR ENERGY**

1   World Energy Council, *2010 Survey of Energy Resources Executive Summary*, 24.
2   REN21. 2011, *Renewables 2011 Global Status Report*, 12, 22, 23.
3   *Modern Marvels: Renewable Energy* (History Channel, 2006), DVD.
4   REN21. 2011, *Renewables 2011 Global Status Report*, 25.
5   Thermo-Dynamics Ltd, "Case Study: Chanterelle Inn, Nova Scotia."
6   MNRE, REEEP, *Solar Water Heating Systems: A Review of Global Experiences*, 12.
7   Sonoma Wine Company, "Solar Cogeneration Installation Unveiled."

CHAPTER FOUR **WIND ENERGY**

1   Archer and Jacobson, "Supplying Baseload Power and Reducing Transmission Requirements," 1701.
2   Dvorak, Archer, and Jacobson, "California offshore wind energy potential," 1.
3   REN21. 2011, *Renewables 2011 Global Status Report*, 19–20.
4   Canadian Wind Energy Association, "Wind Farms: Canadian Wind Farms."
5   REN21. 2011, *Renewables 2011 Global Status Report*, 19–20.
6   Madison County Agriculture, "Fenner."
7   *Modern Marvels: Renewable Energy* (History Channel, 2006), DVD.
8   GWEC, "North America."
9   Sovacool, "Abstract: Contextualizing avian mortality."
10 RSPB, "Wind Farms."
11 Edgecombe Community College, "Wind Turbine installed at EEC."
    REN21. 2011, *Renewables 2011 Global Status Report*, 25.

CHAPTER FIVE EARTH ENERGY

1 Alfe, Gillan, and Price, "Composition and temperature of the Earth's core," 97.
2 REN21. 2011, *Renewables 2011 Global Status Report*, 13.
3 Holm et al., *Geothermal Energy: International Market Update*, 8.
4 Ibid., 7.
5 Pernecker and Uhlig, "Low-Enthalpy Power Generation with ORC-Turbogenerator," 1.
6 Calpine, "Welcome to the Geysers."
7 Canadian Geoexchange Coalition, "Geoexchange General Questions: Is geoexchange used primarily in homes?"
8 Toronto Facilities and Real Estate, "Metro Hall."

CHAPTER SIX HYDRO ENERGY

1 REN21. 2011, *Renewables 2011 Global Status Report*, 13, 71.
2 REN21. 2011, *Renewables 2011 Global Status Report*, 24–25.
3 ROR Power, "Brandywine Creek."
4 Ibid, 55.
5 Alterra Power Corporation. "Projects-Toba Montrose Project."
6 Richard Taylor and David Appleyard cited in REN21. 2011, *Renewables 2011 Global Status Report*, 26.
7 Kanellos, Michael, "A Seventh Way to Generate Power from Water."

CHAPTER SEVEN OCEAN ENERGY

1 EU-OEA, "About Ocean Energy."
2 Pembina Institute, "Energy Source: Tidal Power."
3 Marine Current Turbines Ltd., "Marine Current Turbines secures funding."
4 EREC, "Ocean Energy."
5 OREG, "Ocean Energy."
6 Pembina Institute, "Energy Source: Wave Power."
7 Scotsman, "600 ft 'sea snake.'"
8 Pelamis Wave Power, "CEO at Aguçadoura."

CHAPTER EIGHT BIOMASS ENERGY

1 REN21. 2011, *Renewables 2010 Global Status Report*, 17.
2 United States Energy Information Administration (EIA), "Coal Explained: Where Our Coal Comes From."
3 REN21. 2011, *Renewables 2011 Global Status Report*, 13.

4 RFA, *Climate of Opportunity.*

5 REN21. 2011, *Renewables 2011 Global Status Report*, 32.

6 Westenhaus, "A New One Step Process."

7 CTV News, "Winery scraps."

CHAPTER NINE CLEAN-ENERGY TECHNOLOGIES

1 McKinsey & Company, *Carbon Capture and Storage*, 10.

2 SaskPower, "Boundary Dam."

CHAPTER TEN HOUSEHOLD ENERGY CHOICES

1 AFDC, "Flexible Fuel Vehicles."

CHAPTER ELEVEN COMPARING ALTERNATIVE-ENERGY TYPES

1 REN21. 2011, *Renewables 2011 Global Status Report*, 33.

2 IPCC, "Summary for Policymakers," in *IPCC Special Report on Renewable Energy Sources and Climate Change Mitigation.* 9.

3 Ibid., 6.

BIBLIOGRAPHY

Alfe, D., M.J. Gillan and G.D. Price. "Composition and temperature of the Earth's core constrained by combining ab initio calculations and seismic data." *Earth and Planetary Science Letters* 195 (2002): 91–98. http://www.es.ucl.ac.uk/people/d-price/papers/138.pdf (accessed January 25, 2011).

Alternative Fuels Data Center (AFDC). "Flexible Fuel Vehicles." 2010. http://www.afdc.energy.gov/afdc/vehicles/flexible_fuel.html (accessed January 19, 2011).

Alterra Power Corp."Toba Montrose Power Plant British Columbia." 2011. http://www.magmaenergycorp.com/Theme/Magma/files/assets/_pdf/Canada/Toba%20Montrose%20Power%20Plant%20-%20BC.pdf (accessed July 29, 2011).

Archer, Cristina L., and Mark Z. Jacobson. "Supplying Baseload Power and Reducing Transmission Requirements by Interconnecting Wind Farms." *Journal of Applied Meteorology and Climatology* 46 (November 2007). http://www.stanford.edu/group/efmh/winds/aj07_jamc.pdf (accessed January 7, 2011).

Calpine. "Welcome to the Geysers." 2011. http://www.geysers.com/default.htm (accessed January 12, 2011).

Canadian Geoexchange Coalition. "Geoexchange General Questions: Is geoexchange used primarily in homes?" 2011. http://www.geo-exchange.ca/en/geoexchange_general_questions_faq2.php (accessed January 17, 2011).

Canadian Wind Energy Association. "Wind Farms: Canadian Wind Farms." 2011. http://www.canwea.ca/farms/index_e.php (accessed July 17, 2011).

CTV News. "Winery scraps to help feed electricity venture." November 14, 2007. http://www.ctv.ca/CTVNews/SciTech/20071114/electricity_grapes_071114/ (accessed August 2, 2011).

Dvorak, Michael J., Cristina L. Archer, and Mark Z. Jacobson, "California offshore wind energy potential." *Renewable Energy* (2009). http://www.stanford.edu/group/efmh/jacobson/Articles/I/dvorak-archer-jacobson-2009.pdf (accessed January 8, 2011).

Edgecombe Community College. "Wind Turbine installed at EEC." http://www.edgecombe.edu/news/April_27_10_Windspire_install.htm (accessed July 26, 2011).

European Ocean Energy Association (EU-OEA). "About Ocean Energy." 2010. http://www.eu-oea.com/index.asp?sid=74 (accessed January 6, 2011).

European Renewable Energy Council (EREC). "Ocean Energy." 2011. http://www.erec.org/index.php?id=27 (accessed January 17, 2011).

Global Wind Energy Council (GWEC). "North America." 2011. http://www.gwec.net/index.php?id=24 (accessed January 10, 2011).

Holm, Alison, Leslie Blodgett, Dan Jennejohn and Karl Gawell. *Geothermal Energy: International Market Update.* Geothermal Energy Association, May 2010. http://www.geo-energy.org/pdf/reports/GEA_International_Market_Report_Final_May_2010.pdf (accessed January 13, 2011).

Intergovernmental Panel on Climate Change (IPCC). "Energy Supply." In *Climate Change 2007: Mitigation.* Contribution of Working Group III to the Fourth Assessment Report of the Intergovernmental Panel on Climate Change. B. Metz, O.R. Davidson, P.R. Bosch, R. Dave, L.A. Meyer, eds. Cambridge, UK, and New York: Cambridge University Press, 2007. Also available online at http://www.ipcc.ch/pdf/assessment-report/ar4/wg3/ar4-wg3-chapter4.pdf (accessed January 15, 2011).

————. "Summary for Policymakers." In *IPCC Special Report on Renewable Energy Sources and Climate Change Mitigation.* O. Edenhofer, R. Pichs-Madruga, Y. Sokona, K. Seyboth, P. Matschoss, S. Kadner, T. Zwickel, P. Eickemeier, G. Hansen, S. Schlömer, C. von Stechow, eds. Cambridge, UK, and New York: Cambridge University Press, 2011. Also available online at http://srren.ipcc-wg3.de/report/IPCC_SRREN_SPM (accessed June 22, 2011).

Kanellos, Michael. "A Seventh Way to Generate Power from Water: Pico Hydro." Green Tech Media. April 29, 2010. http://www.greentechmedia. com/articles/read/a-sixth-way-to-generate-power-from-water-pico-hydro/ (accessed January 29, 2011).

Kanter, James. "Carbon Trading: Where Greed is Green." *New York Times*. June 20, 2007. http://www.nytimes.com/2007/06/20/business/ worldbusiness/20iht-money.4.6234700.html (accessed August 2, 2011).

Madison County Agriculture. "Fenner—30 Megawatt Wind Power Generation Facility." http://www.madisoncountyagriculture.com/altenergy/ FennerProjectinfosheet1.pdf.

Marine Current Turbines Ltd. "Marine Current Turbines secures funding to develop new fully submerged Seagen Tidal Turbine & welcomes UK Energy Minister to Strangford Lough. " 2010. http://www.marineturbines.com/3/ news/article/35/ (accessed January 26, 2011).

McKinsey & Company. *Carbon Capture and Storage: Assessing the Economics*. McKinsey Climate Change Initiative, 2008. http://www. mckinsey.com/clientservice/ccsi/pdf/ccs_assessing_the_economics.pdf (accessed January 24, 2011).

Ministry of New and Renewable Energy (MNRE) and Renewable Energy and Energy Efficiency Partnership (REEEP), South Asia Secretariat, *Solar Water Heating Systems: A Review of Global Experiences*. 2010. http://www.aeinetwork.org/reeep/doc/DIREC_Background.pdf (accessed July 14, 2011).

Natural Gas Supply Association. "Natural Gas Supply." 2010. http://www. naturalgas.org/business/analysis.asp (accessed June 14, 2011).

Ocean Renewable Energy Group (OREG). "Ocean Energy." 2011. http://www. oreg.ca/index.php?p=1_7_Ocean-Energy (accessed January 8, 2011).

Pelamis Wave Power. "CEO at Aguçadoura." 2011. http://www.pelamiswave. com/our-projects/agucadoura (accessed January 6, 2011).

Pembina Institute. "Energy Source: Tidal Power." 2011. http://www.pembina.
org/re/sources/tidal (accessed January 21, 2011).

———. "Energy Source: Wave Power." 2011. http://www.pembina.org/re/
sources/wave (accessed January 20, 2011).

Pernecker, Gerhard, and Stephan Uhlig. "Low-Enthalpy Power Generation
with ORC-Turbogenerator, The Altheim Project, Upper Austria." *GHC
Bulletin*. March 2002. http://geoheat.oit.edu/bulletin/bull23-1/art6.pdf
(accessed January 12, 2011).

REN21. 2011. *Renewables 2011 Global Status Report*. Paris: REN21 Secretariat.
http://www.ren21.net/Portals/97/documents/GSR/REN21_GSR2011.pdf
(accessed July 6, 2011).

Renewable Fuels Association (RFA). *Climate of Opportunity: 2010 Ethanol
Industry Outlook*. 2010. http://www.ethanolrfa.org/page/-/objects/pdf/
outlook/RFAoutlook2010_fin.pdf?nocdn=1 (accessed January 22, 2011).

Royal Society for the Protection of Birds (RSPB). "Wind Farms." http://www.
rspb.org.uk/ourwork/policy/windfarms/index.aspx (accessed August 2,
2011).

Run of River (ROR) Power. "Brandywine Creek." 2009. http://www.
runofriverpower.com/id/14 (accessed Jan 25, 2011).

SaskPower. "Boundary Dam Integrated Carbon Capture and Storage
Demonstration Project." http://www.saskpower.com/sustainable_growth/
projects/carbon_capture_storage.shtml (accessed June 19, 2011).

Scotsman. "600 ft 'sea snake' to harness power of Scotland." http://news.
scotsman.com/scotland/600ft-39sea-snake39-to-harness.6303096.jp
(accessed January 6, 2011).

Sonoma Wine Company. "Solar Cogeneration Installation Unveiled."
2010. http://www.sonomawineco.com/renewable_energy.htm
(accessed July 18, 2011).

Sovacool, Benjamin K. "Abstract: Contextualizing avian mortality: A preliminary appraisal of bird and bat fatalities from wind, fossil-fuel, and nuclear electricity." *Energy Policy* 37, issue 6 (June 2009): 2241–2248. http://www.sciencedirect.com (accessed January 15, 2011).

Thermo-Dynamics Ltd. "Case Study: Chanterelle Inn, Nova Scotia." 2010. http://www.thermo-dynamics.com/pdfiles/case_studies/chanterelle_inn/11388_NRCan_Bro_Ev4.pdf (accessed July 15, 2011).

Toronto Facilities and Real Estate. Energy and Waste Management. "Metro Hall—Deep Lake Water Cooling." http://www.toronto.ca/ewmo/pdf/dlwc.pdf (accessed August 2, 2011).

United States Energy Information Administration (EIA). "Coal Explained: Where Our Coal Comes From." 2011. http://www.eia.gov/energyexplained/index.cfm?page=coal_where (accessed July 28, 2011).

———. "U.S. Coal Supply and Demand." 2011. http://www.eia.doe.gov/cneaf/coal/page/special/feature.html (accessed January 18, 2011).

Westenhaus, Brian. "A New One Step Process from Bio-Oil to Petroleum Products." *New Energy and Fuel.* May 15, 2009. http://newenergyandfuel.com/http:/newenergyandfuel/com/2009/05/15/a-new-one-step-process-from-bio-oil-to-petroleum-products/ (accessed January 19, 2011).

World Energy Council. *2010 Survey of Energy Resources Executive Summary.* Also available online at http://www.worldenergy.org/publications/3040.asp. Used by permission of the World Energy Council, London.

INDEX

ACKNOWLEDGEMENTS

Thank you to Martin Clarke for sharing your knowledge and insights. Also thanks to Gary Fabbro for the illustrations, and Lesley Reynolds, Vivian Sinclair and Jacqui Thomas at Heritage House Publishing. My sincere appreciation to all my family and friends for their ongoing support and encouragement.

ABOUT THE AUTHORS

ANNETTE SALIKEN, B.Sc., M.B.A.

As a professional writer, Annette excels at translating complex information into clear, concise, everyday language for a wide range of readers. Her first book, the award-winning *Cocktail Party Guide to Global Warming,* was a bestseller on amazon.ca and has been adopted by many businesses and government organizations as an educational guide on climate change. Annette holds a master's degree in business and sustainability and a bachelor's degree in mathematics and English. Currently, Annette is a senior writer for a large professional-services company. She lives with her husband in Vancouver, British Columbia.

MARTIN G. CLARKE, B.A.Sc., P.ENG.

Martin is a senior professional engineer specializing in alternative energy. In his current role as energy manager at Metro Vancouver, Martin identifies, evaluates and recommends renewable-energy projects to reduce energy consumption and overall carbon emissions from the operations of the Greater Vancouver Regional District. Examples of his initiatives include biomass, micro-hydro and geoexchange energy projects. Previously, Martin was a project manager at a fuel-cell technology company, and a project manager on biomass and waste-energy projects at an engineering consulting firm. Martin is a certified energy manager (CEM) and holds a bachelor's degree in mechanical engineering.

Also by Annette Saliken

cocktail party guide to **GLOBAL WARMING**

Whether you're looking for help sorting facts from sensationalism or simply want to carry on informed discussions about global warming, this straightforward book delievers the goods.

ISBN 978-1-894974-91-2   www.heritagehouse.ca